Water Security

Water Security

Editor by

Judith Rosales

Delve
PUBLISHING

www.delvepublishing.com

Water Security

Judith Rosales

Delve Publishing
2010 Winston Park Drive,
2nd Floor
Oakville, ON L6H 5R7
Canada
www.delvepublishing.com
Tel: 001-289-291-7705
001-905-616-2116
Fax: 001-289-291-7601
Email: orders@arclereducation.com

ISBN: 978-1-77361-007-8 (Hardcover)

ABOUT THE EDITOR

Judith Rosales

Prof. Judith is an Ecologist and Environmental Specialist with more than 30 years of academic and consulting experience. She holds a PhD in geography from the University of Birmingham UK and wrote the PhD and Master Academic Programs of Environmental Sciences for the Universidad de Guayana, Venezuela.

TABLE OF CONTENTS

LIST OF FIGURES

LIST OF TABLES

PREFACE

According to World Health Organization (WHO), water security has been described as "when all people at all times have both physical and economic access to sufficient water to meet their daily needs, and this translates to productive and healthy life". Globally, water is considered as a key physical resource that plays a huge role in improving the pace of general development of a country. Studies have shown that having sustainable water supply system is one way of alleviating poverty and promoting economic growth of a country. Therefore, water security is an essential aspect of livelihood in any part of the world.

Considering the growing world population, water resources are significantly decreasing day by day. It has been estimated that the total world population by 2025 will have grown by additional 2.6 billion from the current population. As the population increases, so does the demand for water. Therefore, without initiative taken to solve various issues on water security in most parts of the world, it is expected in future that more than half of the current world population will lack access to safe and clean water for various daily activities.

At the World Economic Forum Annual Meeting that was held in Davos-Klosters in 2008, different business leaders set out a "Call to Action on Water". Their main goal was to raise awareness in order to develop a better understanding of how water is associated with economic growth across a nexus of issues. This was also to make clear the water security challenges expected to be faced by 2030 especially if a business as usual approach to water management was to be maintained.

Water security challenges experienced in different parts of the world, together with an intensified severity in droughts and flood events which are mainly brought by change in climate are urgent calls for better water risks management. These include management in water shortages, water pollution, and other risks to freshwater systems (lakes, rivers, and aquifers). The main idea is to adopt an appropriate approach that will help in knowing, targeting, and managing various risks and threats of water.

This book contains 10 Chapters that discuss different topics on water security locally and internationally. Some of the key topics discussed under these chapters include introduction to water security, sustainable water supply, Integrated Water Resources Management and poverty reduction, ecosystem and water quality, water pollution, water quality protection, benefits of reusing reclaimed water, harvesting and storage of rainwater, seawater desalination, and solutions to water security challenges in the world.

The issue of maintenance and management of water supply systems remains one of the key aspects of ensuring that there is water security all time. That is why the concept of integrated water resource management has been discussed widely in this book. Another key aspect that has been discussed entirely in this book is access to adequate safe and clean water for all. This is because it is a fundamental human right to have access to clean and safe water for all.

The collection efforts that was put in preparation of this book may not serve as a complete picture of the myriad dimensions of water security, but they reflect some of the difficulties and tensions faced by science in accommodating the idea of water security (based on principles, perspectives, and practice), especially while navigating various economic, political, social, and environmental demand and concerns for sometimes scarce and usually highly variable water resource.

Editor

CHAPTER 1

INTRODUCTION TO WATER SECURITY

BACKGROUND TO WATER SECURITY

With world population now exceeding 7 billion and projected to increase by a billion in every two decades, water security issues are becoming more and more prevalent, particularly in developing countries. Though water issues affect all countries, the rural communities, especially in developing nations are most affected by challenges in water access because they lack proper infrastructure to transport water and may not be in close proximity to water sources. According to World Health Organization view point, 700 million people worldwide face water scarcity and do not have the proper infrastructure to obtain water. Further, this number is only increasing as climate change alters the natural distribution of fresh water around the world. According to current trends, models have shown that over 1.8 billion people could be affected by absolute water scarcity. Pollution also affects the ability of countries to conserve freshwater. In addition to being a requirement of life, water is important because it is vital to the success of agricultural activities. Hence, the inability to maintain water security leads to difficulties in food security as well.

The right of all people to sufficient levels of food suggested by the Universal Declaration of Human Rights is becoming increasingly threatened by the malnourishment and starvation that is affecting large proportions of the populations in developing countries. This problem of hunger that already strikes 925 million people, with the majority living in Asia and Sub-Saharan Africa, will only become worse as time continues and the effects of limited water availability become increasingly pronounced.

Due to water shortages and an uneven distribution of water, certain rural communities in developing countries are having great difficulties in growing food. It has been found that the ability of a community to access water is directly correlated with the poverty and food security of that community. Water scarcity has also been found to lead to gender inequality. Additionally, water scarcity can also lead to desertification, which would cause permanent destruction of important arable land. Hence, water security directly affects food security and both must be protected in order to protect the welfare of communities in developing countries.

BLOC POSITIONS IN WATER SECURITY

Western Block- The Western Bloc, with its more developed countries, will likely help to provide the financial support for international research on improving the efficiency of water use.

Middle Eastern Block- The Middle Eastern Bloc is also more developed than other parts of the world and hence, will play more of a role in funding and investing in the water and food security of underdeveloped countries.

African Bloc- This bloc is severely affected by a lack of water and food security. Hence, this bloc will support international efforts in research and awareness, but must also act on the local level to alleviate the issues of water and food scarcity.

Latin American Bloc- Similar to African bloc, the Latin American also is still developing and will seek the financial support of more developed countries in addition to acting locally to increase water and food security.

Asian Bloc- Though there are many Asian Bloc countries that are highly developed, this bloc also has the largest number of people suffering from water scarcity and hunger. For this reason, policies and actions will differ according to the water and food availability of the respective country.

WATER RESOURCES AND HYDROLOGICAL ASPECTS OF WATER SECURITY

There are three sub-sections of hydrological cycle. These are the subsurface, surface, and atmospheric water systems. Surface water systems comprises of all the water bodies that are directly into contact with the atmosphere such as seas, lakes, streams, rivers, oceans, and ponds among others. On the other hand subsurface water systems comprises of all the water bodies found on the underground and those in the unsaturated zones (including capillary fringe). Most of the fresh waters are mainly stored in the underground water table, permanent snow covers and glaciers. The quantity of water stored in the soil, atmosphere, and various river channels is significantly smaller compared to other water bodies system. Therefore, it is very crucial to understand the water fluxes for human consumption and agricultural use.

Groundwater and river flows are mainly contributed by the rainfalls on the earth that is not withdrawn artificially, transpire, or evaporate. Most of this water on the earth surface is mainly influenced by various factors such as climatic conditions of a place, physiographic condition of a place, the topography (sloppiness and other terrain parameters of a place), river basin size, geological and soil parameters, and the discharge of the river. Most of the activities of human beings on the earth surface play a very crucial role in determining the quality and quantity of water that enter in to the rivers. On the other hand the river can also acts as a carrier or transporter of various artificial and natural components whereby the load is usually determined by the discharge of the river. Therefore, the use of the river water (either for domestic use, industrial use, or irrigation use) is determined by the quality and quantity of water.

To effectively manage various water resources on and underground the earth surface, it is important to first understand all the water cycle processes at different temporal and spatial scales. This in turn helps man to be able to maintain water security. Sometimes this task can be more challenging especially if the interactions that exist between the temporal and spatial variability in water fluxes, water storages, and various residence time components. Additionally, putting the quality of water into consideration, it may become even more difficult to predict the cycle of water especially if human activities are at stake. Based on the figures given in the table below, the renewable periods are usually obtained or calculated based on the mean volume of water divided by the flux mean.

Table 1.1: Illustrating the global water reserves

Location	Volume (10^3 km^3)	% of total volume in hydrosphere	% of fresh water	Volume recycled annually (km^3)	Renewal period (years)
Ocean	1,338,000	96.5		505,000	2,500
Groundwater (gravity and capillary)	23,4001	1.7		16,700	1,400
Predominantly fresh groundwater	10,530	0.76	30.1		
Soil moisture	16.5	0.001	0.05	16,500	1
Glaziers and permanent snow cover:	24,064	1.47	68.7		
* Antarctica	21,600	1.56	61.7		
* Greenland	2,340	0.17	6.68	2,477	9,700
* Arctic islands	83.5	0.006	0.24		
* Mountainous regions	40.6	0.003	0.12	25	1,600
* Ground ice (permafrost)	300	0.022	0.86	30	10,000
Water in lakes:	176.4	0.013	-	10,376	17
* Fresh	91.0	0.007	0.26		
* Salt	85.4	0.006	-		
Marshes and swamps	11.5	0.0008	0.03	2,294	5
River water	2.12	0.0002	0.006	4,300	16 days
Biological water	1.12	0.0001	0.003		-

Water in the Atmosphere	12.9	0.001	0.04	600,000	8 days
Total volume in the hydrosphere	1,386,000	100	-		
Total fresh water	35,0029	2.53	100		

Source: *https://fathertheo.wordpress.com/2012/10/07/water-climate-national-security/*

THE INFLUENCE OF MAN ON HYDROLOGICAL CYCLE

There are several sub-cycle water diversions within the hydrological cycle. These include consumption of water (both by human and animals), drainage cycles, domestic water supply, and irrigation among others. The influence of man directly on the hydrological cycle only accounts for less than one percent of all water resources on earth. This can be depicted in the figure below. However, most of the activities on the earth surface usually play a very significant role in influencing water accessibility and availability of various water services.

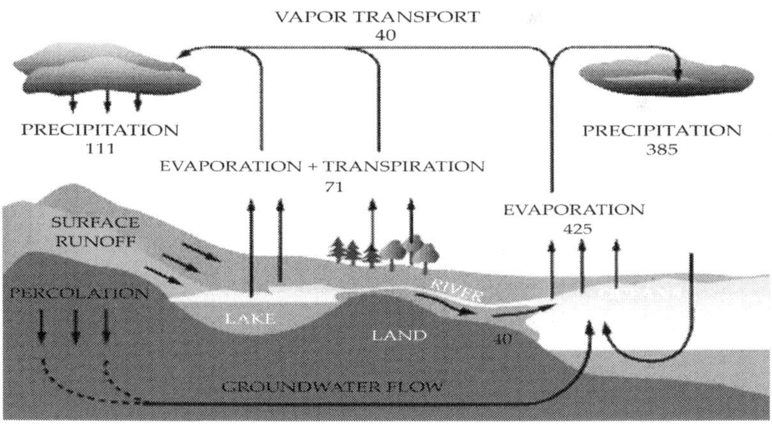

Figure 1.1: Global water cycle schematic sketch.

Source:https://globalchange.umich.edu/globalchange1/current/lectures/kling/water_nitro/water_nitro.html

Due to increasingly over consumption and withdrawal of water, it is expected that availability of water resources will continue to decline gradually almost everywhere on the surface of the earth. The situation is estimated to be worse in various developing countries such as in Asia, African and all those other emerging countries. The following table gives a summary of expected water resources availability in different continents by 2025.

Table 1.2: Projected water use and renewable water resources by continent between 1990-2025

	Available renewable water resources			Water use			
Arca	10^9 m^3/year	m^3/ person in 1990	m'/ person in 2025	1990 with-drawal in 10^9 m^3	1990 consump-lion in 10^9	2025 with-drawal in 10^9 m^3	2025 consump-lion in 10^9
					m^3		M^3
Africa	4,047	6,180	2,460	199	151	331	216
North America	7,770	17,800	12,500	642	225	836	329
South America	12,030	40,600	24,100	152	**91**	257	123
Asia	13,508	3,840	2,350	2,067	1,529	3,104	1,971
Europe	2,900	3,990	3,920	491	183	619	217
Australia and	2,400	85,800	61,400	29	16	40	23
Oceania							
World	42,655	7,800	4,800	3,580	2,196	5,187	2,879

Source: *http://www.nzdl.org/gsdlmod?e=d-00000-00---off-0fnl2%2E2--00-0----0-10-0---0---0direct-10---4-------0-11--11-uk-50---20-preferences---00-0-1-00-0--4----0-0-11-10-0eucZz-kr-00&cl=CL2.9.5&d=HASH0179071ab8456a66f672db1c.5.1&gc=1*

Water scarcity index is one very important indicator used in water security. This index is given by the following equation: water scarcity index= $=\frac{W-S}{Q}$, where W represents withdrawal of water from the cycle by all sectors, S represents desalinated water usage, and Q represents renewable resources of freshwater from the hydrological cycle annually. If water scarcity index is greater than 0.4, then a region is said to be suffering from very hostile water scarcity or stress. It is estimated that more than 2.5 billion people worldwide live in areas with extreme water scarcity.

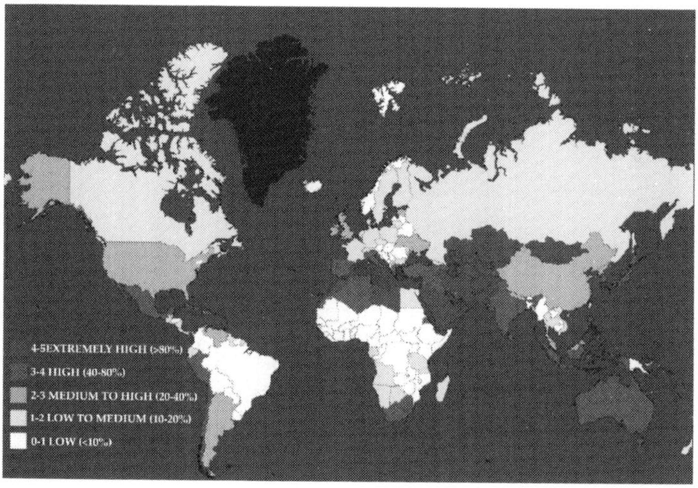

Figure 1.2: Water scarcity index in the world.

Source:http://www.wri.org/blog/2013/12/world%E2%80%99s-36-most-water-stressed-countries

Depending on economics and climatic conditions in the future, this number is estimated to increase to about 3-5 billion people worldwide by the year 2025. Some of the countries suffering from extreme water scarcity or stress include Central Asian, Northern China, Middle East, Pakistan-India border, most parts of Northern and Southern Africa, parts of Europe, and most parts of Central and Western parts of United states. Water scarcity in these areas is projected to be even severe considering the increasing human population, water usage per individual and increasing living standards. Additionally, there are a considerably big number of people who are likely to live in areas that are prone to floods, especially in metropolitan conglomerations. Most of these areas are mainly situated in deltas, river valleys, and along the coastlines. Floods in urban areas may be as a result of excessive rainfall. Alternatively, floods may also be caused by extreme discharge of the river (either due to excessive rainfall, snowmelt or storm surges near the sea).

REAL TIME DATA MONITORING SYSTEM FOR ENSURING GROUND WATER SAFETY AND SECURITY

Ground water is the main source of drinking as well as for irrigation water. Ground water security and safety is of major concern amongst decision makers and all other stake holders. It is therefore important to monitor ground water to ensure its availability in adequate quantity and acceptable quality. Further most of the piped drinking water supply systems are ground water based and functioning of these systems is crucial for ensuring service delivery. Regular monitoring of these systems is also important for ensuring provision of drinking water.

Thus ensuring ground water safety and security is a complex process and involves several parameters and many actors in decision making. Current monitoring system needs strengthening to ensure quick and more accurate response to problems in ensuring ground water safety and security.

In the past decade information and telecommunication technology has grown by leaps and bounds. This technology can be put to use for monitoring different indicators of ground water in a cost effective way, to ensure ground water safety and security. It is proposed to use this technology for establishing a real time monitoring system

The main objective of the proposed real time data monitoring system is to inform the existing decision making system; enabling its functionaries for prompt and accurate actions.

In this document a concept note has been prepared to help in elaborating features of 'Real Time Data Monitoring System' for ensuring ground water security and safety. It gives broad features of the proposed system and thereafter elaborates on proposed indicators for monitoring.

Features of Real Time Monitoring System

As elaborated earlier, the proposed system envisages collecting data of various parameters illustrating condition of ground water and informing the decision making system for appropriate actions. Thus the proposed system is linked to actions for ensuring ground water safety and security.

The system envisages periodic data collection of relevant predetermined indicators illustrating status of ground water safety and security.

Different stakeholders and decision makers will need data in different forms for further actions. Thus the data collected needs to be analysed in different formats to inform the decision making process at various levels.

Currently stakeholders involved in ensuring ground water safety and security include amongst others, at the village level, village water person, functionaries of Drinking water supply department, Medical officer of PHCs , functionaries of health care department and state level functionaries of concerned departments. The monitoring system proposes to inform each of these functionaries with relevant data. Thus the main features of the system are-

- Ease in data collection
- Use of cost effective telecommunication technology for gathering of data collected
- Use of information technology for data analysis
- Informing key decision makers with relevant data
- The system is guided by need for increasing accuracy and speed in actions

Following schematic explains of the system in a nut shell

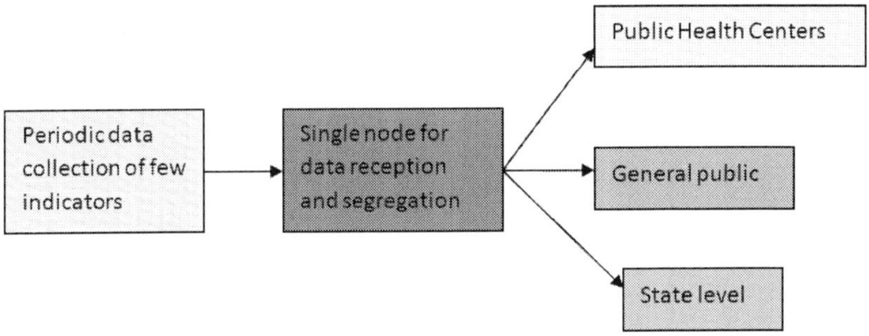

Figure 1.3: Schematic real time data monitoring system for ground water.

Source: *https://sandrp.wordpress.com/2017/02/09/ground-water-2016-indias-water-lifeline-continues-to-bleed/*

Further, it is useful to link this data base with a 'village water safety and security portal'. Analysed data can be displayed for much wider use. This portal can additionally give easy access to other Rural water supply related information/ guidelines etc.

Speed of action can be enhanced by establishing protocols for action. For example, in case of inaction at a lower level for a stipulated time period, the problem should automatically get transferred to higher level of decision making; thereby ensuring speedy action.

IDENTIFICATION OF INDICATORS

As mentioned earlier, ground water safety and security involves, ensuring ground water availability, ensuring its quality and ensuring service delivery. Based on this premise, key indicators can be identified for real time monitoring system. The following paragraphs elaborate these indicators.

Indicators for water security- Ground water is the main source of drinking water and each year ground water resource gets replenished due to rains during wet periods. Currently, most countries assess availability of ground water through a network of 'observation wells'. Each year every country should prepare a 'scarcity management plan' wherein villages likely to face scarcity of drinking water are identified. This identification is in 3 stages and villages that are likely to face scarcity are grouped depending on the period during which it may face scarcity. For example in some locations the first list contains villages that are likely to face scarcity during October to December, the second list villages may face scarcity during January to March and the third list are villages that may face scarcity during April to June. Each region in the world would have different time periods to analyse. Scarcity mitigation actions are guided by these lists. It is possible to refine this process to make it more accurate by monitoring rainfall and water levels in drinking water supply wells. Thus 'rainfall' and 'water level' are proposed surrogate indicators for monitoring water scarcity.

Indicators for water quality- In most countries, owing to Deccan Trap geology, extent of chemical contamination of ground water is less. However, bacteriological contamination is more commonly observed. Regular/ daily chlorination of water is suggested for ensuring potable quality of water. 'OT test' and H_2S test are the most commonly used methods for monitoring chlorination of water. Monitoring kits should also be supplied in order to monitor quality of water with the help of trained personnel. Training to this effect should be imparted to at least one person in every village. In undeveloped countries however it is observed that, effective monitoring is not carried out at the village level as expected. It is therefore proposed to monitor results of 'OT test' and 'H_2S test' through this monitoring system. This will inform the general public for further action. In addition, health care functionaries should also be involved in monitoring quality of water and they should be empowered with the data collected through this monitoring system.

Indicators for Service delivery- Ground water is extracted using either a power pump based piped water supply system (PWSS) or by using hand pumps. Functioning of these systems is crucial for delivery of water supply service. Daily functioning of PWSS and number of HPs in operation will indicate status of service delivery. Operation and maintenance of PWSS is generally the responsibility of the state government and monitoring of daily status of functioning PWSS will ensure efficiency. Technical support for mitigating problems in functioning of a PWSS can be deployed at earliest in order to restore the system. Further non-payment of electricity bills can lead to disruption of service. It is therefore proposed to monitor payment of electricity bill to ensure service delivery. In case of hand pumps, the district machinery can be better informed for prompt maintenance requirement. This will be effective especially in districts where annual maintenance charges for each hand pump are deducted from the country's grants at the district level, and mechanical wing is entrusted with the responsibility of hand pump maintenance. Thus periodic data can be collected for the following indicators by the 'Real time monitoring system'

- Rainfall
- Depth of water level in drinking water supply well
- OT test result
- H_2S test result
- Number of functional hand pumps
- Functioning of PWSS
- Payment of electricity bill

It is now important to elaborate on frequency of data collection; and its link to expected actions.

EFFECTS OF POPULATION GROWTH RATE ON WATER SECURITY

The issue of water security can be summarized based on three categories of countries. These include developed countries, emerging or developing countries, and least developed countries. Some of the developed countries are mainly found in the central and western Europe, North America countries, a few countries in south and Central America, some countries in the Asian region and a larger number of various Oceania countries. On the other hand, the developing countries include countries on eastern parts of Europe such as Russia, some countries in the south and Central America, a larger number of Asian countries such as India, China, and Indonesia, and a larger number of African countries. Lastly, the least developed countries include African countries, several Asian countries, one country in the Central America, and all the smaller Oceania countries.

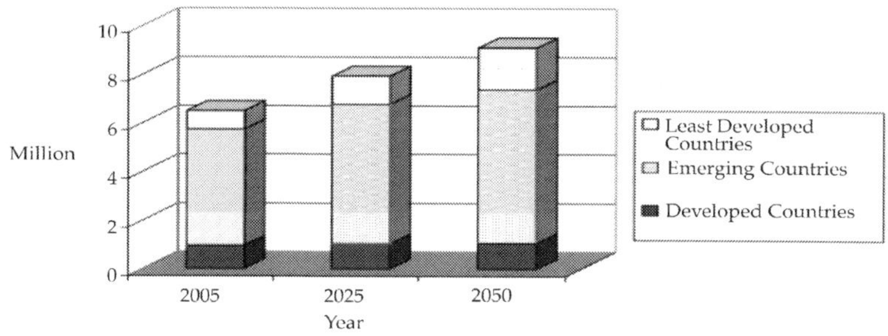

Figure 1.4: Population growth in various developed countries, emerging countries, and least developed countries worldwide.

Source:*http://www.english-online.at/geography/world-population/world-population-growth.htm*

Studies have shown that most people (about 75%) in the world live in developing countries. It has also been revealed that the highest population growth rates are experienced in the emerging countries. Population growth rate in most of the developed countries has been controlled significantly. Although the standard of living is rapidly increasing in most developing countries, it is estimated that about 1.5 billion people in these countries are still suffering from poverty (Gross National Income of less than 1 US$ per day) and more than 70% of them are living in rural areas.

Based on the global scale, the use of water represents a very small percentage of the total water resources. In other words, there is still enough reserve to cater for the future needs. However, because most of the source water is mainly contributed by the river discharge not all that water can be utilized by human beings. Only a small percentage of river runoff water can be used and the remaining water is usually drained into the seas and oceans. Recently, there has been increase awareness on the exploitation and use of groundwater in most parts of the world. Most of the water is used in various agricultural activities and this has contributed to high volumes of water that is lost through evaporation.

Table 1.4: Illustrating global water usage projection in $10^9 m^3$/year

Item	1950		1990		2025	
	withdrawal	consumption	withdrawal	consumption	withdrawal	consumption
Agriculture use	1,124	856	2,412	1,907	3,162	2,377
Industrial use	182	14	681	73	1,106	146
Municipal use	53	14	321	53	645	81
Reservoirs	6		164		275	
Total	1,365	894	3,580	2,196	5,187	2,879
World population in millions			5,176		8,284	
Irrigated area in 10^6 ha	101		243		329	

Source: *https://www.slideshare.net/ING/water-report-2015*

Water needed for agricultural use-Management and usage of water in agriculture is highly influenced by the population of the world, population growth rate, and the general living standards of people. There is an interaction between water management, food production, and living standards of people. These interactions play a very crucial role in influencing the security of water in most part of the world. These three elements usually influence population density of a place compared to the overall area of a country. Based on the table illustrated below, it can be seen that the continent of Asia is leading in terms of both population and population density. This has been referenced based on the total arable land and total area. Putting into consideration the population growth depicted on the figure 1.3 above, one can deduce that most field activities in management of water to improve water security may be expected in Asia in the coming decades. Africa on the other hand comes second in water management activities focussing on increasing awareness on both food and water security.

Table 1.5: Continent and countries categories ranked in terms of population density referenced to the arable land

Continent	Total area in 10^6 ha	Arable land in 10^6 ha	Total population in million	Population density in persons/km^2 with reference to	
				total area	arable land
Asia	3,339	547	3,765	113	688
Africa	3,031	201	840	28	418
Europe	2,299	307	732	32	238
Americas	4,016	3M	850	21	221
Oceania	856	55	31	4	57
world	13,425	1,497	6,215	46	415
Developed countries	3,877	445	1,137	29	255
Emerging countries	7,231	903	4,332	60	480
Least developed countries	2,433	145	750	31	515

Source: *https://www.geolounge.com/continents-population-density/*

Water needed for urban and industrial use-Research has shown that there is a great increment in people migration from rural to urban areas. For instance, in Latin America and the Caribbean, it is expected that more that 85% of people will be living in urban areas by the year 2020. Because of this migration, there has been a growing demand of domestic water due to improved living standards in urban areas. Because of this growing demand, water consumption per capita and water supply demands in urban and industrial areas is swiftly increasing. In most least developed countries, water consumption is estimated at about five litres per person in one day whereas in developed countries water consumption rate per person in one day is estimated around 100-150 litres.

This has significantly affected the usage and requirement of water for sanitation, especially in various urban areas. Due to scarcity of water especially in most least developed and developing countries, water supply in urban and industrial areas has be a big problem and this has impacted hugely on the lives of many people (such as lack of quality water supply and increase cases of water-borne infections). Initiatives have been taken especially by most developed countries in trying to curb the problem of water insecurity and sanitation concerns in urban areas. For instance, European Water Framework Directive is an initiative taken by European Union to control pollution on the surface waters.

Wastewater reuse in agricultural activities has also helped a lot in management of water especially in arid and semi-arid areas. Wastewater recycling has been a very beneficial initiative in managing local water resources. However, there are usually some health concerns that are associated with this reuse if it is not done in the right way.

Sometimes there may be a competition between the agricultural water supply and urban and industrial water supply. This is because urban and industrial water supply is usually supplied at a substantially higher price compared to agricultural water supply.

Water needed for other uses-Apart from agricultural, urban, and industrial use, water is also used for recreation purposes and transportation. In this case, singular provisions are usually required. For instance, large-scale investments may be required for commercial ship transportation (such as in construction, maintenance, and operations) in respect to sluices, canalisation, protection of storm surge, improvement of river bed, and harbour facilities. Most of these investments are usually justified based on the political and commercial considerations. However, consumption of water is usually not direct; rather it is used differently based on the requirements at hand. There are different measures set to improve and promote these water usage types. However, there are some consequences that usually arise on other types of water usage. This means therefore, consequences on other water uses must be considered when providing water for transportation and recreation. These consequences include both socio-economic and environmental implications.

Another very significant use of water is generation of hydroelectric power. According to data obtained based on the International Commission on Large Dams (ICOLD), it has been revealed that there are more than

49,000 large and numerous small dams in the whole world. Most of these dams have been built for different purposes such as generation of hydropower, agricultural purposes, municipal, flood control, industrial and domestic water supply, navigation, sewage water dilution, log transportation, lake water levels stabilization, improving estuaries, and for supplying reservoirs. The main aim on the construction of any dams should be to fulfil most of the above mentioned objectives.

Virtual water trade-This is a concept that is used as a measure to manage and improve water security. This concept was developed in order to explain how countries can relax scarcity of physical water especially in dry regions through importation of water-intensive foodstuffs and other industrial goods. In other words, virtual water trade concept is based on the principle that food trade is used virtually as a mean for water trade. This is because most countries in dry regions can capitalise on importation of foodstuffs and other industrial products and they use their own water resources for other purposes (such as domestic water use). Goods traded usually represent a very small fraction of the total weight of water that is required to produce those goods. Therefore, transportation of produced goods is much easier compared to transportation of water itself. When using this concept, it is highly advisable to consider the fact that the quantity of water required to produce the product does not essentially reflect the quantity of water that will be saved through virtual water import. This is because water that is required will vary regionally because of the variations in terms of water usage efficiency and crop yield.

Statistically, virtual water trade is mainly practised in various countries in dry regions such as Middle East, most countries in east and south East Asia, and North West African countries. Some of their virtual water sources include Oceania countries, North America, and Europe. Normally, water efficiencies and crop yield in most of the exporting countries are usually higher compared to importing countries. Subsequently, exporting countries usually have smaller "real water" compared to the "virtual water" in most of the importing countries. For instance, one kilogram of corn corresponds to about 1.8 tonnes of "real water" in the U.S. and about 2.5 tonnes of "virtual water" in Japan. In other words, virtual water trade of one kilogram of corn from the U.S. to Japan saves about 0.7 tonnes of global water resource.

In the year 2000 the overall virtual water trade (the imported virtual water) for goods was about 1,140 $km^3 y^{-1}$ and this corresponded to about 680 $km^3 y^{-1}$ of the overall water, thus presenting about 460 $km^3 y^{-1}$ water saving. Although virtual water trade cannot improve the overall water resources in a country, "saved water" can be put in to other usage such as domestic and municipal usage. When using this concept in water security, it is highly advisable to be careful because there are some elements such as cultural, social, and various environmental elements that are usually ignored or not considered.

Flood water management and protection-There is an increasingly growing population in various areas prone to flooding (such as along the coastline and in urban centres). This population growing trend is not expected to change in future. Therefore, there is the need to develop strategies on how to protect and manage floods in these areas. One way is to come up with the best scheme that can be used in managing and protecting flood both in urban and rural flood prone areas. The government responsible can also help in creating legislation, policies, key regulations, and other protection works. Other small participants such as local governments, river basin authorities, municipalities, drainage agencies, and farmers may also help in one way or another in controlling flood waters. However, different countries in the world have different approaches on how to handle, manage, protect, and deal with flood water.

Most measures that are taken to manage and protect floods in most parts can be categorized in to two sections. That is;

- Structural measures and
- Non-structural measures

Most of the structural measures are usually tailored toward construction of dikes, dams, barriers, and storm-surge. In other words, these measures are put in to practice in order to reduce flooding risks. On the other hand, non-structural measures are concerned with flood warning, food problems forecasting, flood mapping, plans for evacuation, and zoning land for use among others. Practically, most river basins usually have integrated flood management and protection measures for both urban and rural areas. Development and implementation of both measures as an integrated approach in most effective in managing and protecting flood water.

Most lowland regions especially in the East and South Asian countries have experienced rapid cities growth. For instance, in the last few decades ago huge cities such as Hanoi, Bangkok, Jakarta, Manila, Shanghai, Osaka, Wuhan, and Taipei have received alarming population growth. Some cities are currently holding more than 3 million inhabitants and other even more than 10 million inhabitants. To be able to accommodate these huge population growths, reclamation activities have been experienced in most of the low-lying areas in the neighbourhood of these cities. From a management and flood protection viewpoint, this means that most water storage area have been removed thus increasing discharges of urban drainage. Therefore, creation, implementation, operational, management, and maintenance of integrated measures on flood protection and management can be used to solve various issues expected in these urbanised areas. In other word, development and implementation of management and flop protection measures are required in urbanised areas. For instance, in August 2005 there was a very severe flooding menace in the US New Orleans city that was caused by hurricane Katrina. This flooding can be used to explain the need to properly manage and protect urban cities against floods. Cities along the Orinoco river in South America have been affected by historic floodings during the wet season of 2017.

FUTURE DEVELOPMENTS IMPACTS ON WATER SECURITY

Climate change is currently one of the most important and key influencers of the water security in most parts of the world. For the last few decades there has been several concerns regarding the possible effects of climate change on water security. These concerns include changes in the river regimes, increase drought and dry weather spells, increase in average rainfall per year, a rise of sea level, and effects of design, functioning, and maintenance on water infrastructures.

With prolonged cases of drought-spells, agricultural activities and other industrial activities may severely be affected due to the lack of adequate rainfall and enough water supplies. However, with application of prompt and proper management and flood protection schemes can be beneficial in counteracting impacts of climatic changes in most places.

Population growth in future is also likely to cause impacts on water security in one way or another. For instance, urbanization and population growth rate is likely to cause increase value of both private and public property (such as building, housing, public facilities management, infrastructure maintenance, and other properties). This is going to impact greatly on the decision-making of the management and protection practices and measures on floods.

CHALLENGES AND STRATEGIES TOWARDS WATER SECURITY

For decades, production of food has been relying on various social economic aspects and water access has been one of them. Water is required for both livestock rearing and for growing crops. Adequate water supply is one aspect that should always been observed in order to sustain the livelihood of mankind on planet earth. For the last few decades, maintaining the balance between adequate water supply and water demand for human and livestock consumption has been occasionally been threaten by various elements such as urbanization, growing population, and most recently, climatic weather changes.

One of the significant effects of greenhouse gases build-ups in the atmosphere is likely to cause a significant increase in rainfall variations in most parts of the world. This in turn may intensify the severity of certain conditions such as floods and droughts. For instance, climatic conditions have for long been attributed to devastating flooding in certain regions such as Pakistan. Apart from floods in Pakistan, there has be several cases of mudslides in other areas such as China and India and fire cases in Russia among other cases of climate changes. Variation in rainfall in different places could lead to devastating impacts on food shortage both in urban and in rural areas, thus affecting the livelihoods of most people around the world.

Freshwater is a scarce resource*:* There have been severe cases of freshwater scarcity in most places of the world. Out of the 1.4 km^3 of water in the whole world, only 2.5 percent is freshwater (fit for human being consumption). Most of the freshwater resources are usually inaccessible. This is because more than 70% is in form of ice, snow and stored in glaciers. One of the potential freshwater sources is underground water which accounts for more than 8 million km^3. Only 0.3% of this freshwater (105, 000 km^3) is obtainable from streams, rivers, and the lakes. Availability of freshwater has increasingly aimed at management of water in order to improve water security (which is people's access to adequate, affordable, and safe water in order to satisfy their daily requirements for domestic usage, for food production, and sustainable livelihoods).

Physical scarcity: Physical scarcity can also contribute to water insecurity. Some of the contributing factors towards physical scarcity may include unsustainable overexploitation or consumption, geographical factors, and various climatic conditions. Water insecurity can also arise from poor infrastructure, economic factors, unavailable water resources, and other natural contaminations which prevent people from accessing the water sources. Water insecurity has been a common case in most developing countries in the world. For the last few decades ago, the demand for freshwater in the world has increased by six-folds. About 3.5 billion people in the world (that is about 45% of the total population in the world) live in those places with high demand of water but low water supply. This scenario is projected to become worse in the next few decades as the population growth, industrial and agricultural expand, and economies develop.

Climate changes: Most places of the world are experiencing unprecedented climatic weather changes, with extreme temperatures in some regions. For instance, it has been estimated that the current average temperatures are roughly 0.5 degree Celsius higher than they were in 1990s. This temperature change has really affected most freshwater resources in most parts of the world. For instance, increase in temperature results to a rise in average sea levels by 1.75 mm annually. This has significantly contributed to non-polar glacier retreatment, thus lowering the flow of water in dry-seasons. Increase in sea level has also contributed to increase marine and borehole temperatures.

Trapping of the solar energy plays a very significant role in atmosphere in driving hydrological cycle. Therefore, any alteration caused on the atmospheric greenhouse gases may intensify the hydrological cycle, thus causing a change in rainfall pattern. This in turn may lead to extreme climatic conditions such as floods and droughts. Currently, climate change effects on water security have been felt in most part of the world.

According to IPCC (Intergovernmental Panel of Climate Change) report, the total area classified under "very dry" has doubled since the year 1970. Apart from "very dry" there has also been an increase in cases of flooding especially in mid-high latitude areas, prolonged and frequent drought conditions especially in most of the African and Asian countries, and more intense and frequent El Nino and La Niña occasions. Most of these changes have affected the balance between the supply and demand of freshwater resources in most parts of world.

Vulnerability of developing countries on the impacts of water scarcity due to climatic conditions are partly associated with poor institutional capacity in coping with water supply changes, low incomes, and due to their overreliance on water-based industries such as the agriculture. All local communities and other policymakers at national level in these places should sufficiently anticipate, prepare and adapt for freshwater resource shortages. This will help in countering issues such as starvation for different million people. Water usage and wastage should be addressed amicably. Globally, agricultural water usage accounts for more than 70%. Therefore, in order to optimize on water usage and efficiency, it is crucial to undertake both technical innovations and agronomical research. To address all concerns pertaining water insecurity or scarcity in agricultural activities, revision of various agricultural policies and practices should be undertaken. This will help in improving water sustainability. For instance, all governmental and other trans-national agencies should focus on coordinating the use of water across all sectors.

STRATEGIES APPLICABLE TOWARDS WATER SECURITY

Providing freshwater to everyone in the world is one of the greatest empowerment tools that governmental bodies can use for the high vulnerable and poor communities around the world. However, there are several challenges that face provision of freshwater to everyone in the world. Some of these challenges include poor governance, lack of appropriate knowledge to understand the regulatory framework and policies for implementation, and lack of participatory and transparent processes in water management practices. Most of these challenges can be handled amicably through combining both social and scientific strategies with the help of water national, international, and policy maker experts.

Seminars can also be very useful in sharing and communicating to all people involved about both the traditional and latest knowledge and key practices that should be taken in alleviating issues of water insecurity around the globe. Seminars also help in discussing the key issues and possible solutions towards water problem and what management can do to help in solving some of the issues. Some of the issue that can be discussed in seminars pertaining water insecurity include information dissemination to all water policy-makers, academics, water managers, students and all the other public citizens. There are a number of strategies that can be included in order to address the issue of freshwater insecurity. These strategies are discussed below.

Addressing issues of climate changes: One of the long-term challenge and a threat to many people is climate change. Climate change threatens development for both current and future generations. Climate change usually worsen the issue of water management globally through causing a rise in sea level, melting glacier, variations in amount of rainfall, and other extreme and life threatening events such as drought and floods. There are several consequences associated with climate change. It has been revealed that climate change usually pose challenges on management issue of natural water resources. It also acts as a barrier to transitional process from poverty to prosperity. Therefore, it is important to address an issue to do with these challenges globally and try to come up with possible parameters that can help in adaption of sustainable strategies to these problems.

Allocation and use of resources: Climatic weather condition changes are considered as one of the most life threatening issue towards attaining development and prosperity in a country. Therefore, governments should try to come up with strategies that are applicable in the allocation and use of resources as a mean of curbing water insecurity. There is the need to address issues pertaining competitions in water demand amongst various sectors such as domestic, industry, and agriculture sectors. Regarding this issue, the management involved should come up with possible prospects and mechanisms that will help in addressing water usage, allocation, and sharing of various water regimes. Some of the key issues that can be considered include;

- The current and future water insecurity scenarios both at regional and global levels in conjunction to climate changes
- Possible solutions and mechanisms that can be used in addressing competition and demand of water amongst different sectors
- Joint managerial strategies that can help in addressing various challenges and benefits associated with trans-boundary freshwater regimes
- Other optional strategies and mechanisms that can help in addressing issues of allocation and usage of water resources

Natural water systems: There has been a continuous influence of climate changes on availability of water resources, especially in those dry regions such as South Asia. Climate change has a great impact on the hydrological cycle and natural water system. There are a number of components of natural water resources that affect availability of water resource. These include melting glacier (this mainly affects various patterns of river flow) and variability of rainfall (which mainly impact on the monsoon system, water availability, and rise of sea level). Water management should deliberate on various capacities, knowledge gaps, and limitations that exist to help in prediction and modelling scenarios. These include;

- Present and future scenarios that can be used to predict climate changes and variations in water resources in different regions around the globe
- The status of current glaciers in different regions and their impacts on river capacity and flows
- Bottlenecking and developing scientific and social mechanisms of predicting climate changes and possible options that can be used to address these issues

Water and food security: Most developing and least developed countries are faced with different crises of food insecurity. Food security is one of the key and complex phenomenon that is influenced by different factors ranging from accessibility to utilization of different food products. There are different strategies that can be adopted in order to help in addressing the issue to do with food and water insecurities in the entire world. Those involved should deliberate on possible mechanisms that help in addressing productivities of crop-water as a way to curb the issue of water and food insecurity in the world. There are several perspectives and strategies that can be used in governing water for food security. These include;

- Identifying all known and possible implications that may arise due to climate changes on agricultural water especially in dry regions such as South Asian
- Developing innovative mechanisms and strategies that will help in improving production of crops in case of climate change scenarios
- Addressing different issues that are involved in various irrigation practices imposing on water security
- Developing viable policies that help to address both food and water security issue in dry regions

Science and technology in curbing water insecurity: Both scientific methods and technology can be utilized in addressing water insecurity in most parts of the world. For instance, with the help of viable technological methods, one can enhance efficiency and effective use of water in all sectors. The use of latest technology and other scientific methods can help in addressing and solving issues pertaining water contamination, enhancing efficiency in water utilization, and adopting strategies that can help in water conservation. Additionally, one can use latest technologies to try and explore various ways and avenues required to address and respond to different problems facing many people in water security. Some of the key issue that can be considered include;

- Utilization of all available options for effective and efficient technological intervention to enhance water security in most parts of the world
- Addressing all innovative mechanisms and constraints in developing technologies that are cost effective in water security
- Addressing all the roles and benefits of developing scientific and technological mechanisms to improve water security in different regions of the world

Governance, policies, and regulatory framework: Sustainability in water management is usually defined by governance that is effective and receptive policies. Governance is a multifaceted and holistic framework which incorporates all the economic, social, legal, and political structures. It is vital to develop a policy framework that is responsive and appropriate for governing adaptability and sustainability to climate changes. Some of the key issues that can be considered include:

- Clear identification of stakeholders and levels of action
- Addressing implications of policies to climate changes
- Identifying appropriate measures to innovatively address governance that is adaptive in curbing various issues of water insecurity vulnerability
- Developing appropriate mechanisms (such as climatic-centric mechanisms, mainstreaming climate changes for reformation, and strategic development practices) to address water insecurity
- Prioritizing renewed and responsive policies to address water security based on the changing climate

United Nations (UN) INVOLVEMENT IN WATER SECURITY

On 28 July 2010, the UN General Assembly declared the right for all to clean drinking water. As previously mentioned, the Universal Declaration of Human Rights also affirmed the right of all for equal access to water and food. The United Nations has become increasingly concerned with the inability to maintain these two rights of food and water access as hunger and water scarcity rapidly spread. Hence, in order to combat this, the Food and Agricultural Organization of the United Nations (FAO) publish a report on the State of Land and Water Resources (SOLAW) every 3 to 5 years in order to advocate for this cause. Moreover, the International Conference on Freshwater held in Bohn Germany from December 3rd to December 7th suggested to 118 countries and 47 intergovernmental agencies Recommendations for Action (RFA). The Earth Summit 2002 in Johannesburg also helped to describe the issue of decreasing water availability, urged the implementation of National Sustainable Development Strategies, and helped define the roles, responsibilities, and methods of action of different agencies and governments in water security issues. The United Nations has also supported the research of various international agencies on the issue of limited water and food availability to produce explanations for this trend and ideas for change.

CONCLUSION

Governments of respective countries should develop new and creative solutions that align with their countries policies; here are some solutions that have been previously proposed.

With around 80% of all available freshwater going to irrigation compounded upon the need for water for the ever-increasing global population, it is a matter of outmost importance for water security that water conservation practices in agriculture must be employed. Farmers in developing countries must be made aware of the increasing water demand and limited water supply because they hold the solution to this problem.

To reduce the amount of freshwater used in agriculture, technologies regarding higher yields of crops with the use of less water must be used, and continuously developed and researched.

More developed countries should invest in the water systems of developing countries to increase water productivity and conservation. International research efforts should also be made to allow for more efficient methods of water transport and production.

To promote international awareness, World Water Days and other ideas that are similar in nature and identical in cause to World Water Days should be further promoted. Education about individual water efficiency should also be put in place in both developing and developed countries.

An effective solution will encompass many of the aforementioned methods of attaining water security. It is the extent and the specifics of each method that will help develop the proper response to solve the problems of water scarcity, malnutrition, and starvation in developing countries.

CHAPTER 2

SUSTAINABLE WATER SUPPLY

INTRODUCTION

Sustainable water supply can be described as having sufficient, wholesome, palatable and potable water available within reasonable access. According to World Health Organization (WHO) recommendations, sustainable water supply should be within 50 meters of dwelling and at a minimum of 35 litres per capita on a daily basis. Additionally, this water should be free from all objectionable odours and taste, free from colour and bacteria, and free from chemicals injurious to health. Some of the key sources that are usually considered as safe for water supply include tube wells, piped supplies including public taps, protected water sources (such as springs, dug wells, and ponds), impounding reservoirs, rain water collection, and treated water. Stewardship is one key element of ensuring that there is adequate clean and safe water supply for all in a country. This involves taking responsive actions, engaging different stakeholders, and influencing other for better water supply.

Figure 2.1: *Illustrating stewardship towards enhancing water supply.*

Source: *http://us.anteagroup.com/en-us/blog/how-activate-world-class-water-stewardship-strategy-within-your-company*

IMPROVED WATER SUPPLY

According to the most recent Multiple Indicator Cluster Surveys (2009-2010), the overall access to improved water supply in most parts of the world is 75% in urban areas and 60% in rural areas. However, there is considerable variation between the best and worst served states and regions across the world. For instance, in Asian countries, especially in the Middle East 99 % of the population gets their drinking water from an improved source. However, these coverage figures represent the combined average of rural and urban areas, thus potentially obscure very low rural access. Among the poorest households 70% use an improved water source, while in the richest households the percentage rises to 95%.

If this source type were to be discounted, given the doubts about the bacteriological and salinity quality of such water supplies, this would significantly reduce the level of use of drinking water in most countries around the world.

The highest level of service, household pipe connections are available to 20% of the urban population but only 5% in rural areas. The most common water supply type in urban and rural areas are public tap/stands pipe serving 20% of the urban and 15% of rural population, and rain water harvesting systems for 0.5% urban and 35% of the rural population. Even among rural communities that have been provided with an improved water supply, problems of access continue to occur in the dry season. For example, ponds dry up forcing people to return to less protected sources that are often more distant from the community. In other areas where it has been necessary to drill deep wells and install motorised pumps, these are typically only used in the driest two to three months when other sources are unavailable because communities are unable to afford the fuel and maintenance costs. In coastal and delta regions there have been problems of saline water intrusion in groundwater source. Most governments have addressed this problem with the installation of rainwater roof catchment systems in several rural schools and health centres. Many International Non Governmental Organisations (NGOs) have driven the construction activities of rainwater harvesting tanks for school and community water supply.

ISSUES ON WATER SUPPLY

Microbiological water quality issues are of priority concern in most countries both in rural and urban areas. Although use of improved water supply has increased significantly over the last few decades, the service level is such that it is vulnerable to contamination at source or recontamination during handling, distribution and storage. For example, intermittent piped supplies can experience negative pressure and subsequent contamination. Furthermore, the situation of ageing networks, absence of chlorination facilities and supplies means that recontamination during distribution is quite probable. Surface water and shallow groundwater supplies are also vulnerable to pollution. Communal systems such as public taps/stand pipes and wells necessitate the collection, handling, transport, storage. Unhygienic behaviour of water before end point of use which presents further opportunity for contamination as a result of poor hygiene practices.

Most International NGO's provides insights into typical household practices and knowledge with respect to drinking-water quality. In most parts of the world, water is mostly collected in traditional clay pots, buckets and aluminium pots all of which are prone to hand contamination through wide openings and are uncovered. In the home most households store drinking water in clay containers that are dipped with a cup or beaker to draw water. The most common means of household water treatment is filtering water through a cloth following its collection (55%). Ten percent of world population usually boil water, 3% usually add alum or chlorine, while 39% do not practice any method of treatment. There is clearly potential for stored drinking-water to become contaminated by hand contact when a cup is dipped or a dirty utensil is used. The use of a cloth to filter drinking-water can also lead to microbiological contamination.

Most people understand the link between contaminated water and disease but usually define good quality drinking-water in terms of turbidity, smell and taste. Although microbiological quality is seen as the priority concern, chemical quality issues are also a problem in some parts of the world, especially heavy metals. For example, iron and chloride in groundwater are particular problems in the delta region, leading to a widespread preference for surface water and to a lesser extent rainwater. In most parts of Middle East, salty water intrudes into the groundwater aquifer and surface water sources, a time being declined fresh water in the rivers. In addition there are indications that arsenic and fluoride are present at levels considerably higher than WHO drinking water-quality guidelines and National Drinking-Water Quality Standards of a country. UNICEF has proposed a programme of arsenic mitigation in those townships where high levels of arsenic have been observed. However, an expanded programme of testing and retesting of sites for confirmation of the extent of the arsenic problem has been proposed and funding is currently being sought to support this initiative in those countries that have high risks of arsenic water contamination. Another big challenge and weakness in the management of water quality in is the lack of legislated national drinking-water quality standards in most countries. Different stakeholder agencies (both governmental and non-governmental) are involved in water supply development in order to enhance accesses to water sources that are fit for human consumption.

WATER SUPPLY AND HYGIENE PROMOTION

Most governmental agencies in most countries should work with other international NGOs such as UNICEF on hygiene education in community water supply and sanitation projects. Most of these projects can be evolved into the social mobilisation programme focusing on the training of trainers and advocacy for self-help hygiene promotion such as rural latrine construction and safe hand washing. Hygiene promotion in water supply can be achieved through ensuring that there is adequate clean and safe water, prevent contamination of water sources, proper utilisation and storage, build capacity and create enabling environments to make water safe from contaminants.

Social mobilisation activities in hygiene promotion through water supply involve the school network to further broaden the reach of the programme. Different countries have developed school network projects into a School Sanitation and Hygiene Education (SSHE) programme. This incorporates hygiene education into the 'life skills' curriculum in schools. It also encourages students to observe what happens in their own homes and communities and to practice what they learn in class. The aim is to facilitate child-to-child and child-to-parent communication for improving hygiene behaviour.

Government role in hygiene promotion through water supply should focus more on the training basic health staffs and, community leaders, different volunteers health workers and persons from NGOs in issues of improved water handling, water source protection, hygienic practice on water fetching to water consumption to mouth, household water treatment including chlorine disinfection, filtering, and also in the construction of fly-proof latrines. Additionally, village health volunteers, community leaders and NGO staff should also be involved in training activities.

UNICEF reports that the increase in hand washing after toilet has been negligible, from 46% in 1996 to 48% in 2001. However, the increase in using soap or ash to wash hands after toilet has been significant from 18% to 43% over the same period. UNICEF notes that there is strong correlation between ownership of a sanitary latrine and hand washing with soap after defecation. However, data presented in the Save the Children study suggest that only 13% of households practice hand washing with soap after defecation. After cleaning a child's bottom, hand washing with soap is usually slightly more common at 15%, though nearly a fifth (18%) do not wash their hands at all. Regardless of the apparent discrepancy in practice levels, it is clear much greater attention to motivating hand washing at critical times is needed.

Save the Children has identified another hygiene issue of concern which relates to the disposal of children's faeces. According to research study conducted in Uganda, households with children less than three years of age were asked how they disposed of children's faeces. Forty percent said they threw it outside the yard, 30% leave the excreta on the ground and only 5% dispose of it in a latrine. In households with older children, 3-12 years it was reported that 63% practice open defecation in or around the house. Given the normally high pathogen loads in children's faeces, this area of hygiene behaviour must also be given greater priority. World Health Organization confirms that construction and utilization of sanitary latrine help reduce the diarrhoea in 34%, by use of improved water supply in 36% and hand washing practice help reduce diarrhoea in 40%.

There continues to be a need to maintain a high commitment to hygiene promotion efforts in water supply. It is also necessary that such programmes need to be further scaled-up and better resourced in order to include the rural communities in the more inaccessible parts of the country. It is widely accepted that hygiene behaviour change is a long-term process and will require a concerted effort to achieve the goal of preventing water and sanitation related disease.

INTERNATIONAL ORGANISATIONS AND NGOS INVOLVED IN WATER SUPPLY

Many international organisations are active in the water supply and sanitation sector, especially in most of the developing countries. The United Nation (UN) agencies that are active in the sector are World Health Organisation (WHO), United Nations Children's Fund (UNICEF), and United Nations Development Programme (UNDP). The various programmes of the UN agencies are aimed both at strengthening national institutional capacity as well as direct implementation. For example, WHO is mainly involved in providing technical support to different governmental and non-governmental agencies for water quality monitoring, and producing the comprehensive country-level assessment for water supply. However, WHO focus most efforts on providing support to the Ministries of Health for developing policy and planning, and implementing pilot projects on water supply and sanitation.

UNICEF is closely involved in providing support to social mobilisation activities, developing the Security, Safety, Health and Environment (SSHE) projects. Research shows that UNICEF is involved in the construction of water and sanitation systems for schools and health institutions, and additionally in a few under-served communities around the world. Capacity building of managerial and technical staff, extension workers and community leaders is also an important activity area. UNICEF and WHO are both involved in advocacy aimed at securing commitments from high-level decision-makers for effective policy guidelines and greater commitment of increased human and financial resources to the sector. There are also other international NGOs that are involved in water and sanitation activities as part of their development programmes such as Save the Children, International Committee of the Red Cross (ICRC), Adventist Development and Relief Agency (ADRA), and World Vision among others.

A STRATEGIC ACTION PLAN IN WATER SUPPLY

There are nine key strategies that a government can focus on in order to ensure that there is adequate clean and safe water for all in the country. These strategies are outlined below as follows;

- Creating an evidence-based advocacy for increased access, increased efficiency, and increased equity on water supply in the country. This can be achieved through making sure that there are enough resources for equitable access to safe and clean water supply for the people in the country
- Capacity building of institutional infrastructures and human resources development (such as NGOs, government personnel, and various community representatives). The corresponding

strategic objective in this strategy is to strengthen institutional capacity.

- Promoting health and hygiene education through social mobilization and behaviour change communication. The objective in this strategy is to achieve a strengthened institutional capacity and to have appropriate technologies and practices in water supply
- Promoting application of appropriated technology in water supply. The key objective in this strategy is to have effective sector coordination
- Developing a central management information system for validity and standardization of data in water supply systems. This can also be achieved through having effective sectorial coordination
- Developing mechanisms to assure financial assistance to the outreached community for water supply. Strategic objective in this strategy is to have resources for equitable access to safe and clean water supply
- Community participation and empowerment on self-help basis. This can be achieved through having appropriate technologies and practices
- Linking the national rural development scheme and poverty reduction plans. This can be achieved through having effective sectorial coordination.

The main goal of having a strategic action plan in water supply is to mitigate water burden and various disease associated with unsafe water supply, especially among the young children.

When designing a strategic action plan on water supply, it is advisable to remember that an increased level of water resources is not spread too thinly as this may put at risk the long-term sustainability of the interventions. The plan should be designed to progressively improve the scale and diversity of water supply activities on the assumption that there will be a concomitant increase in the level of water resources.

Based on the current economic and political climate in most countries, water supply is given relatively low priority by national governments, bilateral and multilateral donors, and NGOs. This is something of a paradox given that increased access to water brings significant social and economic benefits, as well as the more widely understood health gains. Indeed, there is a strong argument that ensuring access to safe water supply together with the adoption of good hygiene practices will contribute significantly to all of the Millennium Development Goals as outline in the table below. Therefore, a fundamental activity within the strategic plan is to raise awareness and understanding of the underlying importance of safe water supply and sanitation to development progress.

Table 2.1: Links between water supply and Millenium Development Goals (MDG)

Development goal	Link to water and sanitation
Eradication of poverty and hunger	A lack of water resource management, unsafe drinking water and lack of sanitation are key links in the cycle of food insecurity, poor growth, disease, malnutrition and poverty. Irrigated agriculture provides a large proportion of the world's food and irrigation comprises over 70% of overall water use.
Universal primary education	Diarrhoeal diseases and parasites reduce attendance and attention. Girls often stay away from school unless there are female-only latrines. Time spent collecting water take precedence over school attendance and this burden falls on girls. Teachers are unwilling to live in areas without adequate water and sanitation.
Promotion of gender equality	Women bear the brunt of poor health and the security risks from lack of private sanitation or washing facilities, and the burden of carrying water. Increasing women's roles in decision-making to match their responsibilities, and bringing about a more equitable division of labour are known to help improve water supply, sanitation and hygiene. Demonstrating this can help to improve women's status in other ways.

Reduced child mortality	Diarrhoea causes 2 million deaths per year mostly amongst children.
Improved maternal health	A healthy pregnancy and hygienic labour practices reduce the risk of maternal illness. Hand washing is simple, yet effective.
Combating disease: (HIV, malaria and others)	Of the global burden of disease, 23% is a result of poor environmental health, 75% of which is attributable to diarrhoea. HIV treatment is more effective where clean water and food are available. HIV infected mothers require clean water to make formula milk. Water management reduce opportunities for malaria mosquito breeding sites. Clean water and hygiene are important in reducing a range of parasites including trachoma and guinea worm.
Environmental sustainability	Water resource management is key to environmental sustainability. Water resources are under stress. Public health improvements can address the environmental degradation resulting from urbanisation.
Global partnerships for development	Public, private and civil society partnerships help deliver water and sanitation services to the poor.

Source:*https://unu.edu/publications/articles/quantifying-water-supply-sanitation-and-the-mdgs.html*

Substantial increase in resources for the water supply sector is necessary. However, simply increasing the level of resources is not enough itself. Additional resources must be targeted to address specific problems and inequalities in the sector. It is essential that sound evidence is presented to development agencies and donors (both government and non-government) to back-up the demand for greater resource allocation to the water sector. In the strategic plan it is important that evidence-based advocacy is the best approach to securing additional resources and targeting their use appropriately and equitably. Ensuring that advocacy campaigns are evidence-based will provide a more compelling argument for greater resources to be committed to the sector. The Ministries of Health (MOH) for specific countries should be in a good position to advocate for greater investment in the water sector because they have the mandate to carry out water quality analysis and provision of water facilities to both urban and rural health centres. Although there is both a wealth of experience and data within the MOH that form the evidence base for advocacy, it is not presently being used for this purpose. In spite of the extensive information available there are also information gaps and weaknesses in data quality. Furthermore, the reporting and dissemination of such information tends to move through vertical lines of communication as opposed to laterally across the departments of MOH. It is equally important that information sharing goes beyond government ministries and is used to inform and influence the policy decisions of donors and NGOs in water supply activities.

Although use of water supply and coverage has increased significantly over the last five years in different countries, studies indicate a considerable disparity between States and Regions, and also between rural and urban areas. Reasons for low coverage could include the difficulty of access, remoteness, insufficient personnel and technologies unsuited to specific local conditions. The barriers to increasing coverage in these areas must be identified as the basis for a sound advocacy case.

The full potential to improve health through public water supplies will not be reached unless close attention is paid to drinking-water quality. Several issues have been identified that warrant urgent attention. For instance, there is a clear need to advocate for legislated water quality standards and norms. Another advocacy campaign might focus on the public health benefits from chlorinating urban piped water supplies. Intermittent, leaking piped-water systems are extremely vulnerable to contamination and can lead to serious outbreaks of water-borne disease.

WATER SUPPLY SYSTEMS FOR HEALTH CENTRES AND SCHOOLS

Accelerating water supplies to health centres and schools is necessary, especially in developing countries. It is especially important that safe water supplies are available at rural health centres given the likelihood that some patients will be suffering diarrhoeal disease and other faecal-oral infections. Schooling presents a major opportunity to encourage the early adoption of safe hygiene practices, and is fundamental to the School Sanitation and Hygiene Education programme. The provision of basic sanitation and water supply in schools is essential to successfully motivating a permanent, lifelong change in hygiene behaviour.

Having prioritised the issues on the focus of advocacy campaigns, responsible task-force should then need to identify appropriate communication channels through which the advocacy campaigns can be presented. There is an extensive range of different media that could be used including lobbying, reports, leaflets, posters, fact sheets, and the mass media. The campaigns should be prepared on the basis of careful research and data collection. Any campaigns prepared on water supply should be presented to the thematic group for review and discussion. After making any necessary revisions they can then be put into action. An impact evaluation should follow each advocacy campaign. After thorough discussion and evaluating the water supply campaign, successfully advocating for change should be significantly influenced by the implementation of activities corresponding to the other three strategic objectives. Advocacy evidence should be produced through research and development; institutional strengthening should include the quality of data management and broaden the experience base, and through consultation and cooperation with partners it is probable that further issues requiring policy change may be identified.

Table 2.2: Resources for equitable use of improved water supply

Narrative	Verifiable Indicators	Means of Verification	Risks and Assumptions
Objective Increased resources are made available to ensure greater and more equitable use of improved water supply	Greater resources allocation	Budget documents	Advocacy campaigns successful

Outputs Advocacy task-force established, trained and equipped	Task-force established	Revised job descriptions	Funding available
Evidence-based advocacy campaigns focussed on key water supply implementation	Advocacy campaigns implemented	Advocacy documents	Reliable sources of data
Activities Recruit advocacy task-force	Task-force exists	Task-force personnel	Suitable staff
Designate office space & budget	Advocacy office	Budget documents	Funding available
Procure office equipment	Equipment purchased	Invoices	Funding committed
Train task-force in evidence-based advocacy methodology	Staff trained	Training materials	Motivated staff
Identify and prioritise advocacy issues	Key issues identified	Documentation	Staff competence
Identify appropriate communication channels	Communication means identified	Strategy plan	Appropriate channels exist
Prepare advocacy campaigns and documentation	Campaigns prepared	Documentation	Availability of evidence
Present advocacy campaigns to interagency thematic group for comment, & revise	Campaigns reviewed & revised	Memos with comments	Cooperation from sector partners
Run advocacy campaigns	Campaigns implemented	Activity plan	Necessary resources
Evaluate advocacy campaigns	Evaluation done	Evaluation reports	Staff/consultant time

Source: *http://www.lboro.ac.uk/well/resources/Publications/Briefing%20 Notes/BN11%20Allocating%20Resources.htm*

COORDINATION MECHANISMS IN WATER SUPPLY

An important aspect of enhancing proper water supply is to establish coordination mechanisms between the stakeholders from different sectors and at different levels in order to make investment, management and regulatory decisions and to implement these decisions. There are three types of coordination mechanisms that can be established for water supply in any place. These include; collaborative or co-operative coordination mechanism, water marketing coordination mechanism, hierarchical coordination mechanism. Which type of coordination mechanism is appropriate for a particular situation depends on the frame conditions and on the objectives to be achieved. There are different benefits of establishing coordination mechanisms in water supply activities. These benefits are discussed below.

Improving the flow of information- Coordination mechanisms establishment is essential because it enhances the flow of information in order to enable different actors that are entitled to make better informed decisions on the issues they are responsible for in water supply activities. Relevant types of information include factual information about the hydrological and socio-economic aspects of water resources use, as well as information about the needs, interests and views of the different stakeholders regarding interventions in water resources management.

Coordination of decision-making and implementation- Establishing coordination mechanisms, and improving already existing coordination mechanisms can be expected to lead to better outcomes of strategies that involve different decision-makers and implementing actors. Coordination of decision-making may take the form of "harmonization", or the alignment of decisions that are taken independently by different stakeholders. A more far-reaching form of coordination is achieved by joint decision-making. Depending on how authority is distributed, joint decision-making processes may take different forms (for instance, if authority is equally shared, the stakeholders may negotiate a consensus or practice voting according to defined majority rules. In hierarchical relationships, joint decision-making may imply that lower-level authorities make decisions that are then approved or changed by higher-level authorities, such as in different tiers of government). Different actors may be involved in deciding on different aspects of a joint decision, for example a sector agency may decide on technical aspects, and political decision-making body may give an overall approval. Coordination is not only important for decision-making, but also for implementation. Problems arising during the implementation process may be addressed by setting up a forum where the different stakeholders involved in implementation can meet, discuss problems and negotiate joint solutions concerns issues on water supply activities.

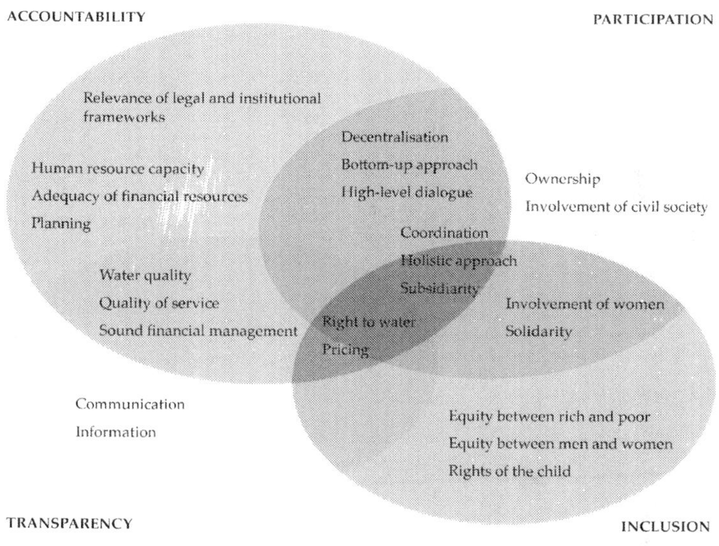

Figure 2.2: Illustrating good governance and coordination mechanisms in water supply.

Source: *http://www.puntosud.org/helpdesk-europeaid/europeaid_at_work/ec_dev_policies_and_sectors/principles_and_objectives/policy_coherence_for_dev/start*

Facilitation of Regulation-Regulatory decisions are type of decisions that often involve challenges of implementation, especially if the institutional capacity of enforcement is low. Enforcing environmental regulations regarding water supply and water resources management is a typical case. Involving different stakeholders into regulatory decision-making (even if there is no legal rule requiring such involvement) may be a strategy to improve the prospects for implementing regulatory decisions. Experience with collaborative management approaches have shown that stakeholders are often more likely to comply with rules if they have been involved in devising those rules.

Promoting Democratic Forms of Governance-The first three reasons for coordination have an instrumental character (they are instruments for achieving better outcomes). Coordination mechanisms that involve citizen participation can also be considered as a goal in their own right, especially in the sense of promoting democratic forms of governance. With regard to citizen participation, one should distinguish direct and representative forms of participation. Direct forms of participation involve, for instance, meetings in which all citizens or all users or potential users of a particular water supply infrastructure participate. Representative forms of participation involve representatives of citizens. These representatives may be the leaders the leaders of community organizations involved in water resources management, or they may be the representatives elected by the general public, such as local government representatives. In traditional systems, traditional authorities may also be considered as representatives.

Apart from distinguishing between direct and indirect representation, one can also distinguish different degrees of participation. Typical classifications distinguish the following levels:

- Sharing information
- Consultation and seeking opinions,
- Sharing of decision-making authority,
- Devolving decision-making authority

Stakeholder participation has become an important concept in water supply and water resources management, and more generally in development practice. Experience has shown that to avoid problems of "token" participation, it is useful that those actors organizing participatory processes ensure transparency as to which type of participation is envisaged, and what the rights and responsibilities of those involved in a participatory process are.

CHAPTER 3

INTEGRATED WATER RESOURCES MANAGEMENT AND WATER SECURITY

INTRODUCTION

This chapter is going to outline and discuss how integrated water resources management practices are implicit in water security issues as they can be used in reducing poverty in the whole world. This is based in the context of land use, public health, livelihoods, food production, urban planning, industrial development, and environmental protection. Specifically, this chapter is going to elaborate on how integrated water resources management can be used within a broader governance outlook to facilitate the decision making process seeking to implement various policies of water and land use. On the other hand the chapter discuss on the connection that exist between poverty and water security and to elaborate on how present approaches in water management can be utilized to reduce the impact of poverty among people; exploring various changes that can be initiated to strategically reduce poverty socially, economically, and equitably; explore integrated water resources management modalities both at national and sub-national levels and discuss their impact on the poverty reduction and water security.

Studies have shown that water access and use are fundamental elements of human survival, productivity, and health. The indivisibility of these functions of water to support human well-being is directly influenced by how the available water resources are governed. Assurance of water sustainability and that of the environment is highly important for the current living generation and the generation to come in future. Integrated water resources management should be done by all institutions (both private and public) either they operate independently or jointly. Studies have shown that this approach of water resources management can sometimes be inappropriate especially because water resources are considered as continuum; hence its usage for a specific purpose can directly or indirectly influence its quality or availability for other usages.

Globally, there has been rapid increase in the competition between people and the use of water resources for different purposes, sometimes leading to massive public outcry and tension internationally. Indiscriminate and uncoordinated usage has also led to serious ground-and surface water pollution and environmental degradation. Having an integrated plan towards management of various water resources can really help in moderating the competition through political participatory, economic and processes mechanisms, and well-informed decision making.

In the competition for access to and use of water, it is necessary to ensure that those with the least leverage in society (many of whom are already without an adequate or reliable water supply or means of waste disposal) do not suffer further disadvantage in the implementation of water-related policies. Studies show that in any competition over access to resources, whether these be natural resources or manmade services and livelihood opportunities, those living in poverty do less well than others, unless they or agents acting on their behalf manage to secure their relative interest compared to those with more economic, social and political clout. For this reason therefore, equity considerations are an important aspect of policy formulation for development interventions; those who are least able to exert their claims should receive special attention.

Figure 3.1: Poor women are the most vulnerable when it comes to water insecurity.

Source: https://www.pinterest.com/pin/185070484451247101/

The chapter does not attempt to argue that an integrated approach to water resources management is a panacea for poverty reduction, rather, the chapter argues that no strategy for poverty reduction can be effective unless water policies in all their dimensions are brought within its competence, and an integrated approach is adopted which allows competitions over access and uses to be fairly and transparently moderated. Although the principle of integrated water resources management has now been widely accepted in the international discourse and to some extent in national water policies, its translation into revamped laws, regulations, institutions and management systems is still in its infancy. However, some case examples already exist which show that integrated approaches to water resources management and use can help people living in poverty to secure their livelihood base (in terms of their survival, health and productivity) and may even help them to improve it.

Therefore it strongly argued that integrated water resources management as a policy tool demands a higher level of significance in the anti-poverty agenda than it is now accorded, and explores the implications for policy-makers of elevating its importance in current international, national, and sub-national poverty reduction initiatives.

ANALYSIS OF WATER SECURITY AND POVERTY

The nature of poverty is a subject about which a vast analytical literature exists, as do theories of causation and response. Poverty is usually defined as "a socio-economic condition in which people's livelihood capacity is inadequate to meet their own and their children's basic needs". This is mainly because of inadequate access to various natural resources such as water. Poverty can also be attributed to lack of unemployment or viable business opportunity. Alternatively, poverty can be defined in the context of some form of special vulnerabilities. Additionally, poverty can be attributed to lack of inadequate access to basic services (such as water supply, health care, sanitation and education). According to World Bank viewpoint, people who endure poverty may define their condition differently, seeing their problems as characterised less by economic disadvantage or lack of service access than by powerlessness, noiselessness, insecurity and fear (socio-political factors). Water poverty can be attributed to;

- Those people whose livelihood dependency on natural water resources is persistently threatened by continuous flooding or drought condition
- Those people whose livelihood dependency on natural water resources is continuously threatened by insufficient water supply
- Those people whose natural livelihood is threatened by degradation, erosion, or a state confiscation (for instance due to major infrastructure construction) without being compensated
- Those people whose livelihood dependency on natural water resource is threatened by long distance to water supply
- Those people whose livelihood dependency on natural water resources is threatened by low household income (for instance, slum dwellers are usually obliged to pay for water well above the market rates).
- Those people whose livelihood dependency on natural water resources is threatened by chemical or biological contamination thus limiting access and use of the water.

- Women and young girls who spend long hours of the day looking for water and whose education, security, nutritional status, and productivity are thereby at risk.
- Those people living in areas with high prevalence of various water-borne diseases such as bilharzia, trachoma, typhoid, cholera, and malaria without any means of protection and control.

Those that are severely affected include the minority groups (especially the indigenous groups), the elderly, those suffering from HIV/AIDS or those other severe infections or mental and physical impairment, those living in informal and shanty settlements, and those who are invisible economically.

Studies have shown that for the last five decades poverty have progressively reduced, but there is still a significant group that still suffer from poverty attributed to poor and inadequate access to water resources. It has been estimated that about 1.4 billion people around the world live on less than $1 per day. Another 2.8 billion people live on less than $2 per day. These statistics have many weaknesses but are the best estimates available of the dimensions of global poverty and 'the poor'.

One or more of the conditions described above mean that the poor endure high levels of infant mortality, child malnutrition, morbidity, illiteracy, and other indicators used as surrogates for poverty. Although the precise interconnections are hard to capture, current patterns of land and water use, and of household and community service provision are integral to the severely disadvantaged situation of the poor within the global community. To clearly understand the connection between integrated water resources management and how it impact on poverty, it is important to discuss these broad statistics rather than on the cited coverage figures that are used to only give information on drinking water supplies and sanitation.

Statistically, those that live on less than $1 per day (about 1.2 billion people in the whole world) directly coincide nearly with the total number of people around the world living with inadequate access to safe and clean drinking water supply (about 1.1 billion people). Additionally, those living with less than $2 per day (about 2.8 billion people) directly coincide nearly with the total number of people living with inadequate access to clean and safe drinking water and sanitation (about 2.4 billion people). Currently, there is no way of knowing whether this symmetry is causal or coincidental but reflects how water security is linked to poverty.

Typically, poverty has occasionally been estimated based on the various economic criteria (using crude wealth-related indicators) such as gross national product per capita. Alternatively, poverty has been attributed arbitrarily levels based on the minimum cost of food basket. Recently, measurement of poverty has been broadened to include all the other social indicators such as infant mortality rate, life expectancy, women status, literacy, access to various services and malnutrition levels. A set of social and economic targets for 2015 covering these areas was agreed at the United Nation Millennium Assembly and Sustainable Development Summit as the defining parameters of a renewed international onslaught against poverty.

- Halve the proportion of people living in extreme poverty
- Halve the proportion of people suffering from hunger
- Halve the proportion of people without access to safe water and basic sanitation
- Enrol all children in primary school
- Achieve universal completion of primary school
- Eliminate gender disparities in primary and secondary education
- Reduce maternal mortality ratios by three-quarters
- Reduce under-five mortality rates by two-thirds
- Halt and begin to reverse the spread of HIV/AIDS

Figure 3.2: Illustrating the Millennium Declaration Goals for 2015.

Source:http://www.eoi.es/blogs/tatianacasquero/2012/01/18/sustainability-analysing-mdgs-progress-in-south-east-asia/

Although the broadening of the debate concerning poverty is usually welcomed in conjunction to access and use of water resources, it is not yet satisfactory. Based on World Summit on Sustainable Development Plan of Implementation, taking steps toward eradication of poverty calls for the "integrated water resources management and water efficiency plans and the promotion of laws that guarantee water use rights, both of which have implications for poverty reduction strategies. But as far as direct poverty linkages are concerned, the principal indicators cited relating to water are still confined to supply coverage of safe drinking water and sanitation services. It is pertinent to note that the major expansion of sanitation services required to meet the target will have to be dry, or 'on site' sanitation. Waterborne sewerage is out of the question for the vast majority of the 2.4 billion currently underserved, on grounds of cost, excessive use of freshwater (for sewage transport), and waterway pollution.

Access and use of Water Resources

There is presently no coherent analysis of the relationship between poverty and water access and use. It may be classified according to water accessibility, to income and living standards, to the satisfaction of 'needs' or to the demand for health, education, electricity and other services. Although water is today acknowledged an 'economic good', its role in economic productivity, ecosystem integrity, and socio-economic status, including that of women, is neglected. The range of poverty analysis relating to water is thus both weak and narrow.

There is a standard assumption that the provision of household water supplies and sanitation will primarily benefit women and children, but analysis of household income levels and food security in relation to water provision is rare. Minor attention has been paid to such aspects as the hours of women's and girls' toil saved from collecting water, but the direct value of water in productive areas of their lives is typically ignored. Only the impact on health has been systematically analysed, to justify household supplies in terms of a reduced burden on health service infrastructures.

Water access and usage is one of the key aspects that directly influence the interaction between the people and their livelihoods. For the majority of the poor people living in developing countries, this consists of rivers, lands, small-scale entrepreneurial activities, forests, and other natural resources. According to an International Fund for Agricultural Development study that was conducted in 1992 it was revealed that more than 2.5 billion people out of the 4 billion that live in 115 developing countries lived in rural areas, of whom more than half lived in those areas suffering from high soil degradations. Another 1.4 billion people have inadequate access to water resource and lived below the poverty line. These people are usually vulnerable to different natural calamities such as drought and flooding which contribute to food shortages. At times of drought condition, even if drinking water supplies remain, the land usually becomes barren, more livestock is lost, and with the exhaustion of their food supply, people are forced to migrate unless relief is provided. The importance of water to livelihoods, as well as to survival, life quality, health, is usually implicit in malnutrition levels, life expectancy rates, poverty rates, and hunger among other social elements (such as urbanisation rates, employment migration, women's status, flood displacements, and school retention rates). Therefore poverty analysis which excludes access to water for purposes other than drinking is highly flawed.

Figures on water use in irrigation and agriculture, and in industrial production and urban services provision, are available but they are not adequately incorporated into poverty analysis – on either side of the balance sheet. Studies have shown that most agricultural and industrial productivity activities are highly dependent on availability and use of water resources. Most governments usually subsidize water for irrigation in most places with an aim of promoting agricultural activities in order to have a steady supply of affordable foodstuffs in a country. Irrigation coverage statistics and crop productivity levels may give indications of national wealth and there are clearly extensive benefits for rural development; however, these may not reach the poor unless they are specifically targeted. The negative social and environmental impacts, such as water-logging and salinity, disruption to downstream fisheries and introduction of water-associated disease are well documented and often affect the poorest first. It is thus essential to determine implications for the poorest members of society of major changes in water use patterns (that is who gains and whether there are losers) in order to reduce socio-economic disparities or inequities.

Water Services Coverage and Equity

The coverage for safe and clean drinking water supplies and sanitation services is considered as one of the key indicators for poverty in most places. However, the existing information and data is inadequate as far as equity considerations to water access and coverage are concerned. Statistics from the Joint WHO/UNICEF Monitoring Programme (JMP) show levels of water supply coverage but do not record whether the installations are functioning. For instance, coverage is clearly lower than statistics suggest in slums and shanty towns where half the urban populations reside; indeed if dwellers are designated as illegal squatters they may vanish entirely from official statistics. According to the International Institute for Environment and Development, data on rural areas coverage are rarely disaggregated to identify the very disadvantaged groups.

Figure 3.3: Water insecurity in shanty towns and slum areas.

Source:http://www.newsweek.com/2014/11/21/water-project-cleans-nairobis-slum-283092.html

In some parts of the world, pressure on the water resource has led to declines in water table levels and serious depletion of quantity because no attention has been paid to aquifer recharge to counteract extraction rates. This has also led to pollution, and to degradation of chemical quality, with the intrusion of fluoride, and heavy metals or salinity into groundwater. It is invariably among the poorest and most vulnerable populations that these problems are most acute, and where there are fewer resources to expand on filtration or other appropriate decontamination measures.

Analysis on Integrated Water Resources Management and Poverty Levels

There is the need to accurately capture the critical significance of water in the protection and provision of livelihood needs. The needs of quality and safe water access and poverty needs to be expanded beyond drinking water and sanitation service provisions. There have been several analyses on various issues related to soil degradation, biodiversity, forests, water quality and quantity in relation to environment costs and provision measures, but the interactions between these phenomena and livelihood systems based primarily on the natural environment have been insufficiently noticed. Where farming and fishing communities have built their livelihoods on the behaviour of natural water courses, lakes, wetlands, and the fertility they provide, the implications of major environmental change for their productivity and household food security need to be factored into analysis. Analysis based on an Integrated Water Resources Management Framework can allow equity considerations to be given a higher level of attention than has usually been the case.

Debatably, in order to give water security its required emphasis based on poverty analysis, there is the need for a paradigm shift on how different people perceive poverty. For several years, the term "water security" has occasionally been used to illustrate or describe the balance that exist between multi-purpose usage of water and water resources sustainability at international level, national levels, community levels, and also at household levels. Water security in these four levels can be monitored based on various indicators (such as qualitative conditions and quantitative conditions) of water resources over a period of time. These indicators should be well-established in all settings as key and integral components for assessing poverty reduction. The first edition of the UN World Water Development Report, to be released in 2003, and the MDG task force established by the UN Secretary-General Kofi Annan after the

World Summit on Sustainable Development provided an opportunity to establish such indicators. In other words, if there is a shift in terms of water and poverty perspectives (which often than not happen if poverty is to be meaningfully addressed) the need for an integrated water resources management approach becomes self-evident.

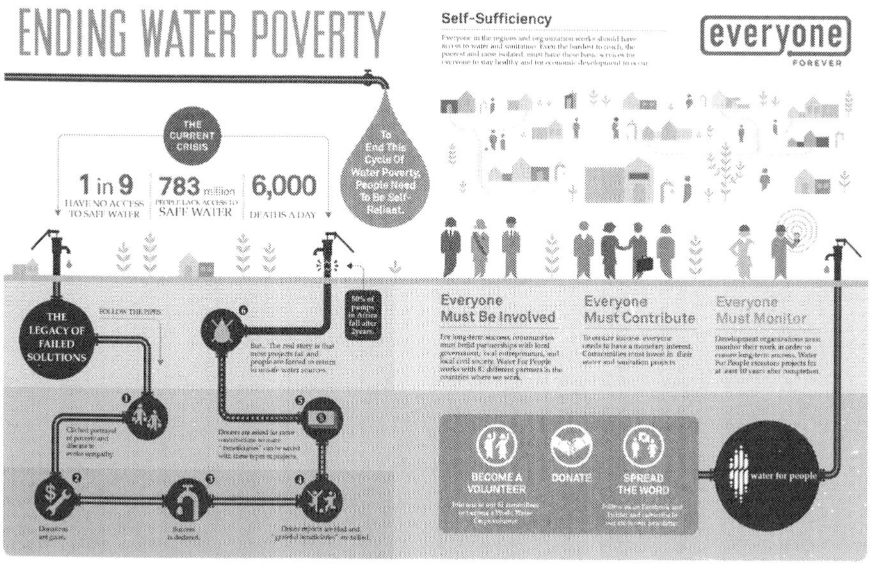

Figure 3.4: A holistic approach towards reducing poverty through water security.

Source:http://www.daviddisalvo.org/the-daily-brain/2013/3/23/how-to-end-the-growing-problem-of-water-poverty-at-a-glance.html

Surprisingly, based on the most recent international concerns on water insecurity, only few governments have been able to highlight water resource conservation as their critical agenda of reducing poverty in the nation. This is a clear indication that there is still lack of awareness of the social and economic importance of water resources base, the need to conserve and protect these resources, and for the sustainability on water service delivery mechanisms among various governmental officials (usually from Ministries of Finance and Ministries of Economic Development), who are mandated with the tasks of preparing these strategies. Therefore, the Water and Poverty Initiative should help in readdressing these omissions.

WATER SECURITY AND POVERTY BEYOND SECTOR-BASED APPROACHES

For many years, water resource management services have been the responsibility of various separate sectorial administrations of different uses of water resource. Drinking water supplies and sanitation are part of public health and/or rural or infrastructural development; irrigation part of agriculture; municipal services under urban administration and public works; and water for mines and industries under industry. Some water-related responsibilities also fall under ministries of environment, transport, fisheries, recreation and tourism. Laws and regulatory regimes have typically been separate for different uses and have not been co-ordinated within one framework and implementation institutions remain unconnected.

Because of these fragmentation on responsibility of water resources management, there has been less concerns on moderating between low-value and high-value competing demands. Lack of information and understanding by different user groups (that is rural/urban, upstream/downstream) about water resource constraints and the need to reconcile competing claims often leads to the settlement of water-related disputes in ways that are politically motivated, unscientific, and short-term. Because of the rapidly growing pressure on water resource worldwide, and the high costs incurred in creating and developing various water infrastructures, it is urgent that we move beyond this fragmented approach. To provide coherence and guarantee the sustainability of the resource and ensure its equitable and efficient use is the genesis of the current push for integrated water resources management.

Sectorial Programmes and Policies on Water Resources and Poverty Reduction

For many years, most governments have consistently focussed on poverty-targeted programmes in rural areas for drinking water supplies and sanitation in order to promote and enhance public health. There have been initiatives to provide water supplies and sanitation in urban areas but only occasionally have these included slums and shanty-towns. During the International Water Decade different low-cost water supplies and sanitation approaches were developed and promoted. These included providing hand water pumps and other stand water pipes in communities and "on-site" facilities (such as pit latrines). There is little information on how poor household as a sub-set have been reached based on water supplies and sanitation. It goes without saying that most poor households in rural societies are now able to access to clean and safe water supply in most parts of the world compared to how it was before. However, sanitation still lags very far behind.

The following case study is a good example on how water supply beyond sectorial approach can really help in solving poverty issues. In 1997, Mayniland Water Services Inc., a consortium formed by Benpres Holdings Corporation and Ondeo Services, won a 25 years concession for water and wastewater services for the West Zone of Metro Manila. Hundreds of pockets of 'blighted areas', containing 35% of households, were deprived of water services, and obliged to buy highly-priced water from private vendors. In 1999, a survey showed that for the poorest inhabitants, those who had no connection to services paid more than twice the proportion of their household income for water, and consumed less than one third of the amount, compared to those who had one. The company determined to provide 'water for all'. For blighted areas, they offered alternatives: public faucets operated by Community Water Associations, or individual connections. Because of their poor experience with public faucets and prospective savings on water expenditure, people preferred household connections. Accordingly a Bayan-Tubig – Water for the Community – project was started. Much of the supply-line was laid above ground, against the walls along narrow alleyways. Batteries of meters were placed in the most convenient location, with plastic pipe connections leading from one meter to each house. This system allowed 58,000 additional household connections to be made, metered and charged for in places where dwellings were rudimentary and even illegal.

Studies have shown that access to clean and safe water supply can directly influence the level of poverty on people. For instance, a survey conducted in Nicaragua on 5,000 different households revealed that about 20-50% of the households with a high income was considerably well-managed compared to the poorest households. More than 40% of additional income came from small livestock and house gardening which were mainly managed by women around their houses. A similar study conducted in Ghana revealed that most poor households practising farming got their extra income through informal irrigation practices of horticultural crops for various local markets. It artificially to state that water source from a household perspective is entirely for personal hygiene, drinking, cooking, and other domestic usages; alternatively, it is also used for other agricultural practices. Demand for water among the poor people is generally used for survival and for household purposes, and this is one aspect that has often been neglected especially by different governments when providing resources with the aim of reducing poverty both at national and community levels. Most people living in various marginalized environments usually view a water resource as holistic element in their livelihoods. In other words, they usually operate naturally based on the principles of integrated water resources management, not minding about the sectorial context in which the resource is provided to them.

Figure 3.5: Water services coverage and equity is a key issue in poor and marginalized community.

Source:*https://practicalaction.org/blog/where-we-work/kenya/poverty-and-the-water-energy-nexus/*

On a large-scale, most governments usually postulate irrigation as a primary element to help in promoting economic growth through increasing levels of crop production (either for food or for cash). This holistically helps in improving national food stock availability and raising income from agricultural export. There have been multiple studies conducted to support this evidence that irrigation actually helps in promoting development and eradication of poverty at community and national levels. For instance, without various irrigation schemes and projects in some countries such as Pakistan and Egypt, their economy would collapse, thus resulting to increased political instabilities and urban migrations. However, the government should address equity consideration on water resources at macro-policy levels. It is important to also address irrigation investment implications among the various households living in abject poverty, as opposed to just generalizing their implications on overall national economic productivity. This is one of the aspects that are frequently ignored. The need to rectify inequities in terms of who gains and who loses have been left to social welfare, food subsidies, and resettlement or land compensation programmes. These have often proved inadequate, reinforcing rather than offsetting economic marginalisation.

There is also enough evidence that policies on agricultural use of land have helped in provision of incentives to increase earnings based on water-incentive cash crops (such as sugar-canes and rice) even in those areas suffering from water shortage. In most countries (both developed and developing), governments are used to providing subsidies to farmers for surface irrigation and also electricity that is mainly used to pump the water used in irrigation. However, sometimes these subsidies may result to water resources wastage and excess pollution due to the wrong economic signal that is usually send to the consumers. This promotes unsustainable water resources management.

Therefore, there is the need for a more integrated approach toward water resources management especially among the poor in order to enhance agriculture irrigation. For instance, investment in creating water resource infrastructures in India and North East Brazil has really helps in creating dynamic economic growth in rural communities. This has helped in improving agriculture reduction and other informal employment. Therefore, most governments have recognized the conventional approach "construction as any cost" is not viable at all in managing water resources. The government has a responsibility of improving the existing water supply systems in order to address financial maintenance and sustainability and identify viable approaches that can be used in reducing water usage. One of the best approaches would be to adopt a water management approach that is less fragmented and more holistic. It is also essential to consider bringing all social, economic, and environmental aspects of water management into a single policy framework.

Water Value based on Demand and Supply

For a long time, clean and safe water has always been considered as one of the key social goods, however, since the 1992's International Conference on Water and Environment held in Dublin, it has also been considered as one of the economic goods. Based on this consideration, there has been a decline for water supply-driven policies by the international support. From the World Bank viewpoint, there has been consistent water wastage and inadequate efficiency from various construction schemes from which the costs of water incurred are usually not covered by the capital obtained from various consumers and these cannot be maintained by all means. There has also been a lot of negligence on proper water supply system maintenance and management and this has really impacted on the supply of water both at national and community levels. For instance, in Indian rural areas, more that 40% of hand pumps and other water supplying systems have run out of action. More cases of water pipe leakages and other illegal take-off have also been reported.

Figure 3.6: Farmers with capital resources have shifted their farming practices into water-intensive farming.

Source: *http://oteplicah.com/kommunikacii/poliw/28-kapelnyj-poliv-v-teplice-svoimi-rukami*

The funds, institutional and regulatory capacity for water management services have for a long time been absent of poorly applied. In most parts of the world, lack of any perception of the opportunity cost of water in alternative uses (a perception which has begun to grow only since water became recognised as a scarce commodity) has resulted to low tariffs for both industrial and domestic users.

Politically, promoting water demand-driven and realistically priced water supply systems has been so difficult, considering the fact that water has for a long period of time considered as an infinite resource both in rural and urban populations. Most farmers with capital resources have shifted their farming practices into water-intensive farming especially in those areas with dry weather conditions. This has led to uncontrollably extraction and depletion of groundwater. Additionally, heavy pollution loads in most of the fresh water lakes and rivers have increased seriously. Many water bodies, particularly during low-flow summer months, are usually heavily polluted. This has resulted to serious epidemic and other public health concerns to member of society. Realistically, there is the need to enhance water pollution regimes and other enforcement measures of water quality both at household and community level.

Demand-driven water supply services can be used as a way of reducing poverty through allowing people to express and cater for their needs. People's "willingness to pay" for adequate water supply services and to mobilise around an essential water resource has been widely been demonstrated in most parts of the world. For instance, an experimental project known as Yacupaj, meaning 'for the water' was started in the Bolivian Department of Potosi in 1991. Up to this time, the assumption had been that farmers on the high mountains "Altiplano" lived such a marginal existence that they could not pay for water or sanitation services. Yacupaj set out to turn this premise on its head. Project implementation rules, technological design criteria, financial policies, and maintenance structures were developed. To participate, groups of local communities had to accept financial and technological constraints, agree to undertake O&M, and find around 50% of the total investment. Although many would have preferred household connections, they mainly went for gravity-flow and public standpipes because that was what they could afford. By 1994, Yacupaj had provided services for one-third of the 115,000 people eligible to participate. Most of those who had not done so did not want to (they were happy at present with what they already had). A survey a year later showed that 90% of water installations and 82% of sanitation facilities were functioning. A national World Bank-assisted $46 million programme grew out of Yacupaj.

However, it is recommended that introduction of a water management approach not to be executed in a way that abandons an earlier government responsibility that is acknowledged to offer poorer sections of the community. If not, users should be willing to pay a price that is more realistic. For the last few decades there has been a rapid growth on water resource management emphases at community levels. This has mainly been focused on water facilities ownership for both community water supply and irrigation schemes. Drinking water supply system management at community level has also received mixed results based on various variables such as skills transfer, technical support, and back-ups from the management. Policies which appear to be oriented towards poverty reduction by decentralising responsibilities and ownership of water-related installations may have considerable benefits if handled carefully but can have the opposite effect if government simply divests itself of costly responsibilities for purely fiscal reasons.

The recognition of water as an economic good requires that water resources are allocated between high and low value uses. However, equating willingness to pay with value is problematic and there are enormous complexities in establishing and managing pricing regimes and developing efficient, transparent and accountable collection systems and the creation of appropriate water right systems. Equitable water distribution through licensing and regulation policies has also received considerable difficulties, especially because those with political and economic clout usually tend to appropriate resources. Sometimes water allocation based on licensing and price adjudication has also been perceived as a way of benefiting the rich people in countries with inefficient and corrupt administrations.

Integrated water resources management cannot be used to solve such underlying socio-political water issues in countries. However, it can be used to provide a viable framework and policies that govern the entire process of water supply system (with equity as a key policy driver) as well as necessary bodies can help in enforcing these policies and framework. Based on water demand-driven approaches, integrated water resources management can be used in allowing the communities and other responsible administrations to proper manage water resources through demand-responsive and participatory (democratic) processes. Lastly, an integrated water resource management approach can help to mediate competing uses and ensure minimal water supply services for both urban and rural poor are recognised as key and high-value participants in designing water pricing regimes.

INTEGRATED WATER RESOURCES MANAGEMENT IN ACTION

Integrated water resource management approach in most parts of the world have been recognised as self-evident in international circles, and IWRM principles are increasingly being adopted within water-related policies and good governance. Gradually, countries are introducing reforms in law, policy and administration which reflect the new concerns in order to enhance efficiency, equity, and sustainability. However, some countries around the world are still lagging behind both economically and politically on matter to do with water security. Reforms on water equity especially in countries where opinion on climate is not handled in the right manner are usually difficult to develop and implement. Most of the time there is opposition to the idea of water service provision. In some settings, politicians have built careers on promises of free or all-but free services for water, power, and other public utilities, and cite a pro-poor agenda to defend policies which are unsustainable and which therefore ultimately cannot serve poor people. In this case, progress towards attaining efficiency on water reform is usually slow. Lack of significant and wide-scale changes among both the political and middle-upper income groups is a big challenge towards attaining better results in water security and water resource management.

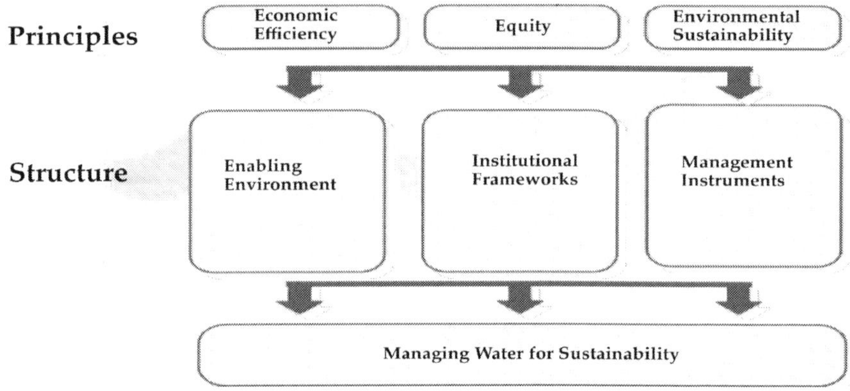

Figure 3.7: An integrated water resources management perspective.

Source:*https://www.slideshare.net/iwlpcu/integrated-water-resources-management-in-malaysia*

However, civil society does not always wait for government to act. There are a number of settings where the reality of water scarcity and its disastrous impact on health status, human survival, and productivity has contributed immensely towards the emergence of local integrated water resource management solutions even if they were not consciously launched under an IWRM banner. The introduction of government-led sectorial programmes during the development era has weakened such community-led solutions and the control of water supplies by impersonal and remote bodies has not always been to the advantage of low-income communities. On the other hand, local level decisions can have adverse impacts elsewhere within a basin and a government led holistic approach is the only means to secure basin-wide sustainable resource use.

Regeneration and Protection of Watershed

Integration towards protecting various water resources can be done in different way. These include through competing use integration, between sectorial integration and through demand of different groups integration (such as all those at risk and those that assume control over water resources when they are given half the chance). Balancing demands requires a political process and negotiation among the parties, whose common interest lies in effective water management.

Aquifer depletion in a place can be as a result of water scarcity or prolonged drought condition. Regenerating watersheds through the use of various tradition technologies is one of the best ways forward towards water security. Different NGOs are also playing a crucial role in mitigating various issues of water insecurity through micro-planning, management of watersheds locally, through provision of integrated water services, and promoting proper water uses among the people of the community. This has really helped in practically dealing with various water problems on the ground. In some places, most of these efforts have been backed by various official policy and support for the development of models provided by external donor agencies. For instance, India and other South Asian countries are some of the places where such an initiative has been practised for a long time. Many cases of community-based ecological regeneration based on rainwater collection and storage (a centuries-old range of hydraulic technologies practised in most parts of the world) have been eclipsed by different governments as a way of supplying rural water services free to communities since the 1960s.

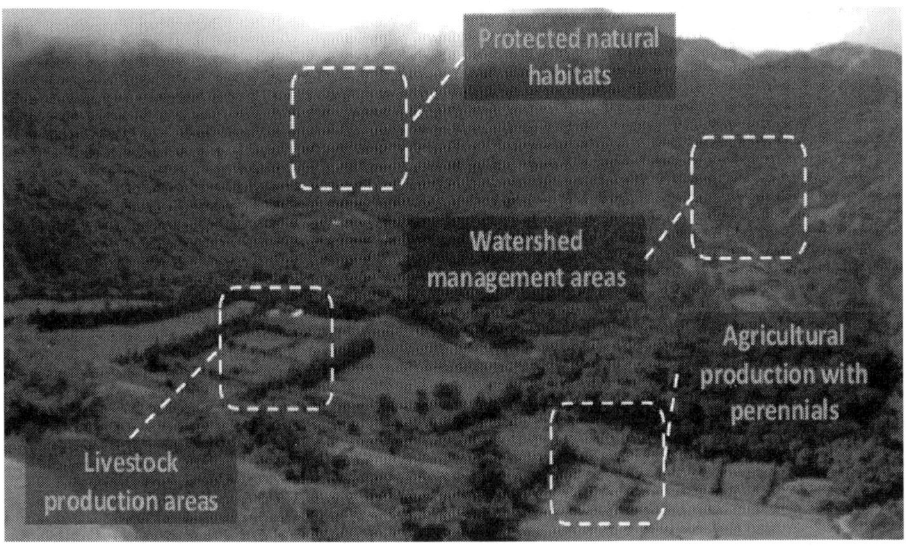

Figure 3.8: Natural regeneration in landscape approaches.

Source:*https://www.slideshare.net/Ecoagriculture/natural-regeneration-in-the-context-of-integrated-landscape-management*

In the face of severe drought, there have been several initiatives of enhancing water security such as construction of communal check-dams and other small structure used for capturing and storage of rainwater run-offs. These structures have for a long period of time played a significant role in recharging aquifers and transforming local eco-systems as well as their surrounding economies. Participatory and transparency in decision making and community rights establishment on prioritising water resources usage have also played critical roles in ensuring there is water security in communities. Local bodies decide priorities, resolve conflicts, and set burden- and benefit-sharing rules. Small rainwater structures are swift to show regenerative results, and within a few seasons the rise in the water table can be remarkable. Results for many communities have been nothing short of spectacular, with continued cultivation and relative prosperity when surrounding areas with identical rainfall are unproductive, their wells dry, and their inhabitants forced to migrate elsewhere.

The following is an example of a case study of a pioneering organisation called Tarun Bharat Sangh (TBS) in Rajasthan. Since the late 1980s TBS has helped communities build 4,500 small dams and other rainwater harvesting structures, transforming the economy of 147 villages. In one area, a 'Water Parliament' (Sansad) was established among the 72 communities along the banks of a section of the Arvari River. Around 300 johads (small dams and bunds) were built by the communities, whose representatives meet in the Sansad once a month to settle issues. Their rules include criticisms on water pumping, a ban on water-intensive crops such as sugarcane and wheat, another on tree-felling, fines for infringements and larger fines for not reporting infringements. However, it is worth noting that 377 cases have been taken out against the TBS for building illegal structures under legislation left over from the British raj, such is the level of obstruction which unconventional initiatives on water and land use can confront.

Examples exist of watershed protection and regeneration in other parts of the world. An interesting case in South America concerns the management of river La Quebrada Chocho watershed where people have been encouraged to protect the land around the watershed through reforestation. This civil society initiative has shown how common projects that have been agreed upon can be implemented successfully among various water users to help in hydrological regeneration of an area. The promotion of various catchment areas and river basins as the logical geographic unit for the realisation of integrated water resources management in practice has attracted much attention, but at levels above the localised, river basins rarely confirm to administrative or political boundaries and there are many examples of tensions between various users between the downstream and upstream in different jurisdictions. In the Guiana Shield, Andean, Amazonian regions, indigenous communities from different watersheds have a large knowledge to transfer about watershed management and water security issues.

Community Management of Water Services

Issues surrounding integrated management of water services include the need to address service sustainability, environmental protection, and equitable provision of supplies for disadvantaged populations. During the past two decades there have been numerous initiatives in various informal settlements in urban, rural, and riverine areas. These initiatives have helped in developing water supply systems of community ownership and water services management in order to integrate water demand and supply. Additionally, these initiatives at community level have also promoted different technologies which are environmentally and economically sustainable. NGOs have also played crucial roles in acting as intermediaries between administration of government and various communities to help work out various water supply systems and structures (financial and technological). Other responsibilities played by NGOs at community levels include fund training and organising different community workers, enabling various associations on water usage, and establishing local water committees to help in water management services. Socio-political dynamics pose many, often underrated, problems in all community-based participatory approaches and such initiatives are not a panacea. More independent evaluation is needed to determine the reasons for success and to determine the means to replicate those successes.

Figure 3.9: Community involvement in management of water resources and services.

Source: *http://lavsodisha.org/watershed.html*

Participation of women is one of the key factors that help in promoting success in management of water services in community. Demand-responsiveness, long-term operations, information sharing, and proper maintenance are some of other key factors that can be emphasized to help in promoting water resource management. Studies have shown that technology choice and community participatory role is also critical aspects in influencing successful outcomes. Most community members are usually aware of their financial and technical limitations. When people are offered with options, most of them would choose a protected water resource in preference of other water supply systems such as hand pumps and electrically pumped supplies.

For instance, In Niassa province of Mozambique, WaterAid has been supporting district administrations and local NGOs to implement a demand-responsive approach to community water supplies. The methodology took into account communities' ability to pay installation and recurrent costs, and they chose from a menu of technologies. Demand rose from nine installations in the first year to 144 in the third. By far the majority of these low-income communities have chosen protected wells as opposed to hand-pumps which they do not believe they can afford. This has released the programme from the standard problem of spending its time repairing and resuscitating non-functional hand pump-boreholes, and while not ideal, protected wells are a major improvement on streams and swamps, the sources used before. There is constant demand from new villages to be included and steadily rising consumption rates, and the outstanding question is how the water committees can adjust tariffs to control demand as pressure on the resource mounts in future years (a reminder that integrated water resources management is a dynamic process).

Many other similar examples of small-scale community water services management exist in a number of countries around the world, providing the point that such initiatives can result to higher service provision levels and productivity for disadvantaged populations. However, it is important to note that most community initiatives are highly influenced by local enthusiasm and commitment on a small-scale. What may be practicable with regard to integrated water resources management at a community level (in a small river valley for example) can be very difficult to 'scale up' significantly since the number of competing interests exponentially multiplies.

A rare example of large-scale integrated water resources management based on watersheds exists in Jiangxi province, China with a population of 41.4 million, which is hilly and mountainous, and suffered severe environmental degradation in the 1960s and 1970s. There was soil erosion, frequent droughts and floods, and accelerating desertification, pollution, and spread of schistosomiasis. In the 1980s, a programme to reverse environmental decline began. It was based on the concept that mountains, rivers and lake (MRL) comprise an integrated system, whose deterioration was leading to poverty. The Jiangxi provincial government set up an MRL Development Office, with sub-branches, to plan poverty reduction programmes based on catchments rather than on administrative boundaries. The emphases were;

- Sustainable development of small watersheds in mountain zones;
- Integrated development of hilly land;
- Eco-agriculture and

- Schistosomiasis control around Poyang Lake

A network of projects was established, supported by 27 demonstration sites and 127 extension stations. The programme, which is exemplar of IWRM principles, has proved that a balance can be struck between environmental protection, economic development and social stability. However, political support of a kind almost unique to China was essential for such a wide-ranging programme in so large a demographic, geographical and administrative unit.

Where governments have begun to decentralise the management of water-related installations and services to local communities with support from appropriate tiers of technical and administrative bodies, many questions remain about responsibilities at the various levels. Most governments around the world are reasonably hesitant in conferring various roles in management of watershed in communities or in large social units in an integrated manner. Issues on integration based on administrative and geo-political structures have been real. Decentralisation of water services has also been perceived as a very risky process to the technical standards and consistency. Additionally, there are has been several challenges in terms of resistance from different allies of bureaucracies, which have hindered implementation of water supply services delivery, and this has led to numerous financial implications. The approach of integrated water resources management may thus be universally accepted, but there is a lot that need to be done in terms of translating this approach into a wider operational practice and policy framework.

Some watershed communities in rural environments have lost faith in the authorities' promises to provide water from macro-schemes remote from their control, which may be dependent on the whims of upstream users, remote administrations and conflicting jurisdictions. In some cases they have taken matters into their own hands. Such ad hoc development however can be a disaster for others within the same basin. For instance, there have been different conflicts between poor communities over water resources along Pangani River in Tanzania which after drying up in its lower catchment. These conflicts have been due to poor community planning on various developments that are initiated. Gold mining in South America is an issue to water security as upstream mining use of the watershed produce negative effects in the waters downstream. As a government, it is therefore significant to promote integration in water resources management through addressing various practical implications that may arise from unresolved issue of water equity. This integration should be done at levels above the localised community. One way of integrated water resources management is to come up with laws and policies that will help in solving various water difficulties at community level. Decentralisation and community-based solutions can also help the poor and marginalised groups but should not become the new mantra as this could lead to anarchy. Based on integrated water resources management framework, the government has the role of allocating and regulating water resources. This can be done through abdicating its responsibilities to various local groups and NGOs. The signs are that politically, the tide in favour of integrated water resources management is turning and a better balance between centralised and decentralised management process can be attainable.

Equity Consideration in Integrated Water Resource Management Policies

When introducing an integrated water resources management approach in the community, proper plan must be done in altering various laws, policies, and other regulatory frameworks in water services. One of the requirements when executing these changes is equity, especially for all the people living in abject poverty. Most disadvantaged groups in society can be affected severely (both positively and negatively) in number of ways by legislations set aside to govern the rights over water supply services. Some of the key protective or affirmative legislative elements over water services include;

- Creating special laws and rights of water access (especially for the people designated as "below poverty line" delinked from land-ownership)
- Protecting both customary and traditional water rights (especially those that are enjoyed by minority and indigenous populations
- Reserving adequate water resources for environmental protection and which should be enjoyed as free good in order to help in sustaining various population groups (especially those with little or no purchasing power)

Recently, both Zimbabwe and South Africa passed legislations regarding water and land use. For a long, the South African National Water Act and the Water Services Act, both passed in 1997/8, are widely regarded as pioneering in their assertion of the "right of all citizens" to access to a minimal level of water provision and held up as a model of integrated water resources management principles. In Zimbabwe, a revised Water Act has attempted to cater for equity in allocation by reducing water rights/permits that were held in perpetuity; and removal of date system concept on water prioritizing and granting of water permits. These provisions are designed to protect the interests of the less well-off (smallholders and farm labourers) who were previously hindered from adequate access due to the extensive granting of water rights and permits to large, predominantly white, farmers. It remains to be seen whether this will in practice lead to more equitable distribution.

Studies have shown that policies alterations can immensely impact on various opportunities for the poor. For instance, there was a new law that was enacted in Mexico 1992 to govern water services. This law brought about several changes on how water supply systems were managed before. This was also focussed on giving poor water users much greater say when it comes to management and conservation issues. In other places different laws have also been enforced in order to bring effects on reductions of water pumping activities from the aquifers. These are initiative taken by different governments to help in conserving and protecting natural water resources. However, within different water policies and laws that have been intended to improve efficiency and conservation, there is the need to careful consider equity. Efficiency in water usage and allocation can be enhanced through water marketing, but unless it is complemented with suitable provisions for various vulnerable groups and for environment protection, water marketing can promote exclusion and negligence.

Similarly, this can be observed if water services are authorized by private sectors where water provisions to various customers who cannot afford paying for the water services and those that use small quantities are neglected because are considered insignificant in terms of viable customers. Studies have also shown that most poor citizens considered as weak and politically unviable in terms of manipulating public utilities are most often excluded in decision-making process regarding water services both at community and national level. Therefore, the only time that the poor will benefit from water services (either publicly or privately owned) is when laws and regulations will be passed to govern equity and sustainability in water resources allocation and management.

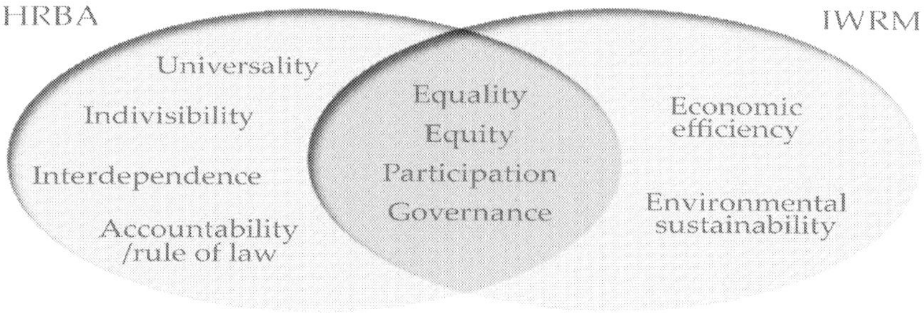

Figure 3.10: Equity and sustainability in water resources allocation and management.

Source: *http://www.waterlex.org/waterlex-toolkit/how-to-articulate-the-human-right-to-water-and-sanitation-and-integrated-water-resources-management/*

In most countries around the world, most regulations and laws that were passed to govern water and land use are still on books. Most of these laws and regulations are occasionally neglected when it comes to application to deal with issues of disadvantaged populations and customary rights deprivation over various natural resources. Therefore, efforts are required in order to reconcile most of these laws and regulations equitably within legal frameworks that are modernized. Otherwise, most vulnerable and marginalised groups who are not able to defend their legal rights on access to various natural resources equitably will continue suffering from their comfort zones. Considerable care is nevertheless required in order to avoid "capture" by powerful groups.

Institutional Reforms and Development

Integrated water resource management approach should ensure that there is equity management and institutional water supply systems applications. Based on the range of activities listed in the figure below, there are two categories of action that can be distinguished. The first category includes those which are tailored towards various deprived groups (including all the interventions aligned towards vulnerable, at-risk, and low-income groups). Additionally, they also include all the changes executed in management and design of services, mitigation of drought, and protection of environment especially for the degraded areas and polluted waterways on which marginalized people depend. The second category includes the initiation of various rigorous focuses in the existing water supply services and water resources management regimes. Based on the various recent emphases that have been directed towards considering water resources as an essential economic good, there have been challenges to successfully implement various market mechanisms in order to help in controlling demand and ensure financial sustainability.

- Ensuring Equity in Integrated Water Resources Management
- The collection of data relating to 'water poverty' which enables the understanding of water-related poverty issues to be expanded beyond service coverage;
- A specific focus on populations 'at water risk': those vulnerable to drought, water scarcity, and flood;
- Orientation of those professionally involved in providing and running services concerning the specific needs of socio-economic* vulnerable groups, including women, children, those affected by HIV/AIDS or other similar condition, indigenous and minority groups;
- Capacity building of professionals in the full range of low-cost and appropriate technologies for water, irrigation and sanitation;
- Capacity-building at all levels for democratic and demand-responsive approaches, espe¬cially at the comet pity level to which local planning of resource use, and management and maintenance of services, will increasingly be devolved;
- A special emphasis on the involvement of women in commmity management mecha-nisms and in policy development;
- Development of partnerships between different sectors, and different types of imple-menting bodies —public, private, NGO — to secure communality of interests and synthe¬sis of pro-poor policies;
- Advocacy of pro-poor approaches towards subsidies and

pricing mechanisms which can be used to reduce inequitable allocations of service benefits and access,

- The active use of existing assessment and data-collecting tools that enable disadvantage, discrimination and demand patterns to be captured (environmental impact assessment, gender assessment, social impact assessment, participatory rural appraisal);
- A stronger policy emphasis on information, education and commmications campaigns to inform people of the links between water, sanitation and health, and to promote hand washing and other key elements of a hygienic way of ***life;***
- The use of IWRM mamgement tools for efficient and equitable management change (GWP, 2002).

Figure 3.11: Equity in integrated water resources management.

Source: *http://trysophal.blogspot.co.ke/*

The introduction of market values within an integrated water resources management framework allows prices to be set for high-value and low-value uses. The difficulty is that, in market terms, a high-value use might well be one in which water is extensively used for production of crops (or raw materials extraction) which commands a high export price, but which exhaust the water resource base. Equally, if a high-value use is assigned to securing a minimum water supply for people living in poverty, it will be difficult to procure resonance in market values. The management of prices, tariffs, subsidies, and the systems needed to implement them in such a way as to ensure equity without intrusion from politically and economically powerful groups, require a high degree of sophisticated management, and a regulatory and operational regime beyond the capacity of many existing administrations. It is not practicable to lay down a set of universal principles concerning the use or non-use of subsidies, or for tariff structures and pricing regimes. The many economic parameters, including users' capacity to pay and the possibilities for cross-subsidies, have to be addressed in each setting. Thus, capacity building for undertaking such exercises should be a priority; and likely shortcomings in implementation should be recognised and avoided.

Since, within integrated water resources management approach, a wide variety of economic, social and environmental factors need to be brought within an overall policy framework, the building of partnerships is an essential part of the political processes required. Links between user groups, service delivery groups, local councils, sectorial bodies, government departments, non-governmental organisations, external support agencies and private companies can yield cost savings, reduce conflicts, and enable the comparative advantages of players to be harnessed. Some of these players have a more authentic understanding of poverty dynamics, and a better track record in addressing poverty issues: some kinds of user groups, local councils, and NGOs are primarily concerned with poverty reduction. But these organisations rarely have broad command over policy-making, nor do they have the resources to implement service delivery or environmental protection programmes on any but a much localised scale.

Bridges need to be built between those bodies which interact directly with water-needy, disadvantaged and vulnerable groups, and those developing or implementing policies and programmes at national level. For instance, in 1998, a Bulgarian NGO (the Centre for Social Practices in Sofia) initiated the Varbitsa Watershed Council to bring all the communities dependent on the Varbitsa River around the table and settle inter-cultural and inter-ethnic differences. The aim was a river-wide development plan, in which local municipalities would sink their differences. In time, it was hoped that this pilot council would enable new approaches to be adopted by the Ministry of Environment leading to larger Watershed Councils. In the first stage of the council's work, conducted under the chairmanship of the Mayor of Kirkovo, an ethnically mixed Muslim municipality, scientists mapped out the river's problems. The Varbitsa is now the first thoroughly researched river in Bulgaria. By 1999, the council had an action plan, and a number of local, inter-ethnic, problems had been already solved by the collaboration of local authorities. The government has now agreed to produce a new Water Law promoting the establishment of Watershed Councils.

The effective use of assessment tools in the planning and implementation of water service schemes and major irrigation and hydropower projects can help the processes of allocation, environmental mitigation, and moderating competing claims in such a way that the interests of equity are served. The social, economic and environmental uses of water by different population groups (such as upstream/

downstream, urban/rural, male/female, rich/poor) and the capacity of available sources in terms of quantity and quality need to be fully understood. Tools such as Participatory Rural Assessment, Environmental Impact Assessment, Social Impact Assessment, Resource Assessment, and Gender Assessment can be used. Data should be shared with user groups, and public campaigns of information and education undertaken to familiarise all members of society with the constraints on water resources and the need for sustainability. Ultimately, it is essential to involve all the stakeholders (both formal and informal) in guaranteeing all vulnerable groups their legal rights in accessing clean and safe water.

An ecological economics approach is needed for evaluating water security in watersheds as it includes also the analysis of perceptions and values of water resources for inhabitants.

CONCLUSIONS AND RECOMMENDATIONS

There has been tremendous milestone in equity, sustainable, and efficiency of water management since the approach of integrated water resource management was introduced and implemented in most countries across the world water pollution and scarcity has also been managed especially in highly polluted and over-populated places. However, some gaps still are existing in appreciation of and implementation of various policies, laws, and regulations, which acceptance of the principles of integration necessitates. There are also several challenges and complexities involved in effectively putting all contexts and levels into effect (technological, managerial, political, behavioural, and administrative). Most of the issues on competition to freshwater access are more felt by some groups than other. Most of these issues can be moderated through the use of integrated water resources management approach. However, applicability of this approach is further complicated by the 1.2 billion poor and highly vulnerable people around the world that need their interest protected. Therefore, it is the responsibility of governments together with other stakeholders to try and come up with viable strategies on how to effectively use integrated water resource management framework to help in making these prospects more achievable.

The sequence and pace at which integrated water resources management reforms are applied are very critical. Water governance can really help in backing up application of these reforms in ensuring that there is water security across the globe. Water governance helps in ensuring that there is coherent in water supply systems. This in turn requires human management and development at institutional level in managing water resources and other water-related services according to a range of social, economic, environmental, and technological principles. These have yet to be fully absorbed by professionals, bureaucrats, and politicians for whom long-established mind-sets and rules of practice are understandably difficult to replace. Not only those specifically involved in water management need to be re-oriented.

Appreciation of roles in water resource management should be emphasized across board, including all financial officials and other economic planners. This will help in dealing with various issues of water pollution and scarcity and their impacts on the livelihood on people. The role of water in reduction of poverty across the globe has occasionally been ignored by different development practitioners and other decision-makers. Therefore, a proactive analysis is needed to assess the various interconnections that exist between water access and other water-related services. This analysis will also help in prioritizing water needs for those people that are marginalized and highly vulnerable.

There are usually losers and winners in any competition involving access to resource. Integrated water resource management approach should therefore not be considered as a silver bullet, especially when it comes to avoiding issues to concerning difficult trade-offs in water services. In other words, it should be used as an approach of identifying and moderating a framework that most stakeholders and users can account on. For poor to have a say in this framework, there is the need for affirmative actions. It is important to also consider the fact that implementation of integrated water resources management will experience some challenges such as political resistance. Therefore, IWRM framework will have to undergo a process of grappling with other trade-offs between theoretical and feasible ideals. Some of the recommendation that emerge from this discuss based on integrated water resource management and poverty reduction include;

- There is the need to emphasize on various policies, laws, and other managerial and institutional structures that are designed to promote the integration of various water services, especially those that promote equity and reduction of poverty. The need for democracy, transparency, and information sharing in the context of major infrastructure projects, especially those which involve substantial environmental changes, should be underlined. Advocacy relating to integrated water resource management should emphasize its value in poverty reduction strategies, which should repair their neglect of water resources management as a necessary component.

- There should be a shift on water resource management from a sectorial approach. This should not only be done based on the unsustainability, inefficiency, and lack of environmental protection, but also in a way that promote equity in water resource allocation and use. Up to now, the only sectorial programmes specifically targeted to poor people have been rural drinking water and (dry) sanitation programmes, and some smallholder irrigation services, conservation schemes and urban supplies projects. However, since their thrust has mainly been to expand coverage, information about their impact on poorer groups is sketchy and assumption-led. Bringing various sectorial policies into one single framework would be beneficial in allocations of water resources between the higher-income and lower-income earners. This will help in promoting equity, democracy, and transparency in resource sharing and access.
- Care needs to be taken that recent changes in water-related policies deriving from the Dublin principles (such as water is a scarce resource and an economic good) not to discriminate poor people. In any competition over scarce resources and services for which payment is required, those with least economic and political clout are likely to lose out unless their interests are defended. Studies have shown that reduction of subsidies, charging of fees, privatisation of services, and decentralisation of maintenance and ownership to community groups, can discriminate against poor people unless introduced in an enlightened, efficient and equitable manner. Integrated water resources management cannot guarantee that this will happen, but it makes the prospects more likely.

- Those cases where integrated water resources management already operates at community and area level should be given a greater degree of formal recognition, and those authorities which have been unwilling to incorporate these approaches into mainstream water- and land-use policies should be encouraged to do so. There is the need to emphasize on efforts towards exploring and solving various issues associated with introduction of watershed-based and catchment-based management for different natural water resources on which most people depend upon. This should be done in co-operation with existing administrative and political jurisdictions.
- Because "water poverty" is considered as one of the key components that contributed towards poverty among many people across the world, and inadequate access to various water resources among the vast majority of the poor, a concerted effort should be made to promote a paradigm shift in thinking about poverty. If the parameters of water's role in poverty were extended beyond drinking water supplies and sanitation to land and water use in productivity and livelihoods, the case for integrated water resource management as part of poverty reduction would become self-evident.

Integrated water resources management framework cannot be used a solution in reduction of poverty, as it has been repeatedly stated across this chapter. However, it can be put in to use in managing and conserving various water resources and other related water services in a way that help in poverty reduction. Any alteration on various laws, administrative structures, and policies that have been proposed may have different implications on both losers and winners. Most of these implications are usually not clear during the inception stages of integrated water resources management framework. Since there are prospects that are usually associated with integration of water resources management framework, it is advisable to always ensure that these opportunities for more productive and healthier lives among different people at-risk and various deprived population groups are not lost, rather they should be transformed into certainty.

CHAPTER 4

ECOSYSTEM AND WATER QUALITY RELATES TO WATER SECURITY

INTRODUCTION

Water quality has been an issue for particular ecosystem types and this has occasionally posed some challenges to many people across the world. This chapter has been prepared to describe various issues of water quality, it also discuss some case studies and other conceptual models to assist water managers in understanding the general water quality and water quantity relationship in rivers, reservoirs, estuaries, wetlands and near-shore marine environments around the world.

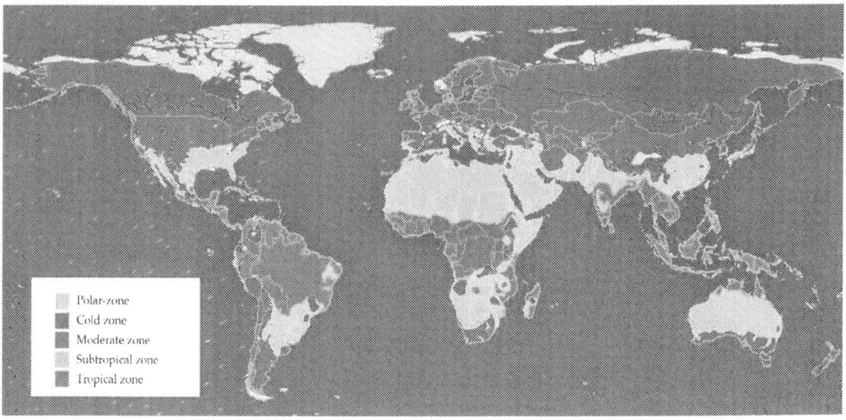

Figure 4.1: Climate Zones of the world.

Source:http://printable-maps.blogspot.co.ke/2008/09/climate-regions-world-map.html

The relationships between the quality of water and quantity of water usually vary for different types of aquatic ecosystems (e.g. rivers, estuaries, lakes, groundwater systems, wetlands and near-shore environments) and vary across the world. There are several factors that contribute to these differences such as geological characteristics of the catchment areas, the topography of a place, and climate. The nature of water flow or watering regimes on the other hand can be affected by alterations of water extraction, land use around catchment areas, and various river regulations enacted to govern the source. Valuable environmental water allocations are available in some systems, for example Commonwealth environmental water holdings, where managers have an opportunity, and a significant responsibility, to create environmental benefits by delivering water to these ecosystems at most appropriate times and in the most appropriate ways.

A good understanding of how water quality interacts with flow/volumes is very useful when making decisions regarding water management. Some water quality indicators respond almost immediately to environmental change, particularly in response to changes in flow or water volume. Also, a decline in water quality is often the first sign that ecosystems are under stress, e.g. hypoxic black-water events in regulated systems, algal blooms in stagnating pools or acid-generation when acid-sulphate soils are exposed then rewetted. In these situations, understanding the water quantity and quality relationships is essential to guide management responses in a timely manner. It is essential for water security management at the local, regional, national and international levels.

One of the main ecological factors related to water security have to do with climate change as it affect the amount and quality of water resources, therefore it is an issue that need to be incorporated in every analysis. During extreme weather conditions such as floods and drought, water management becomes very critical in ensuring that the quality and quantity of water is not altered in any way. Studies have shown that during these extreme weather conditions that the quality of water is usually affected negatively. For instance, during extreme drought weather conditions, the flow of most rivers is usually low or may cease in some, especially the seasonal rivers. This provide favourable conditions for toxins evapo-concentrations, water temperature increment, algal blooming, and low levels of oxygen dissolved. Additionally, during drought weather conditions, contaminants coming from various point sources (such as industrial releases, sewage treatment plants, irrigation returns, and other effluents from dairy facilities) may increase due to low river flows and this usually degrade the quality of water discharged from the rivers.

During flooding on the other hand, there is usually high rate of soil erosion and diffusion of runoffs in different catchment areas. Other common conditions during flooding include sewage overflowing and contaminants mobilisation from various source points. This usually increases suspension of various solid materials, nutrients, and other types of contaminants into estuaries, waterways, and near-shore marine environments.

NATURAL RIVERS

This section describes how natural hydrological regimes affect water quality in different river systems across the world. Most countries have different seasonal rainfall zones and their variability is usually reflected in the flow regimes of rivers across the country. Some part of the country (such as along the rainforest) may have wet winters and drier summers, which results in high river flows in winter and low flows in summer. The Antarctic Regions experience the opposite, with wet summers and drier winters. Semi-arid or arid regions are particularly well represented in most parts of the world, (75% of most continents' land area being classified as having low rainfall). Most dry-land rivers are usually temporary, which means that they do not have surface water flow all year around. This section will describe a temperate (wet winter/dry summer) system, a tropical summer-dominated river system and a dry-land river system.

Temperate Rivers

Most countries have a temperate climate according to the Köppen Climate Classification. The climate is generally mild, ranging from warm to cool, with four seasons (spring, summer, autumn and winter) evident in the weather patterns in these areas.

The temperate regions have the most fertile soils and supports high levels of agriculture. They also support the vast majority of world's population in the major areas. As a result, temperate rivers have experienced high levels of regulation and urbanisation, which has disturbed the relationships between water quality and flow and leading to water quality issues.

This scenario describes the relationship between flow and water quality for unmodified, unregulated temperate rivers to set the baseline for how such rivers would naturally function. Temperate regions can have varying rainfall patterns (e.g. winter or summer dominated), however, this scenario focuses on the regions where the majority of rainfall occurs in the cooler winter months, followed by a drier summer. The flow regime of these rivers follows the seasons with high flows in winter/spring and low flows during summer/autumn.

Water quality, as well as important ecological and geomorphological functions of a river, depends on the magnitude, duration, frequency and seasonality of flows. Unmodified, unregulated temperate rivers (with wet winters/dry summers) typically have some or all of the following flow components;

- Summer cease flow-(which is characterised by lack of discernible flow in the river), However, not all temperate rivers experience cease-to-flow-event.

- Summer base flow or low flow- (whereby a small river flow usually provides a continuous flow through a channel). This flow is mainly derived from the groundwater
- Summer freshes- (whereby small and short duration peak river flow events that usually exceed the base-flow usually last for at least several days often in response to localised storms
- Winter high flow-(usually characterised by persistent increase in the seasonal base-flow that remain within the channel)
- Winter bank-full flow-(whereby channels are usually completely filled with little flow spilling onto the floodplains)
- Winter overbank flows-(characterised by flows that are usually greater than the bank-full flow and normally result in surface flow on the floodplain habitats)
- Water quality is intrinsically linked to each of the above listed river flow components.

Low flows occur over the warmer summer/autumn months. Groundwater inputs often support these flows, but can be saline. The ecosystem is most productive under these flow conditions, with higher rates of nutrient and carbon cycling. Low turbidity levels and slower flow rates allow photosynthesis by aquatic plants and attached algae within the river channel, which produces bioavailable organic carbon to support food webs. Water quality issues such as excessive plant growth are limited by the low nutrient concentrations and shading from an intact riparian zone. This functioning relationship between flow and water quality is described using the Riverine Productivity Model.

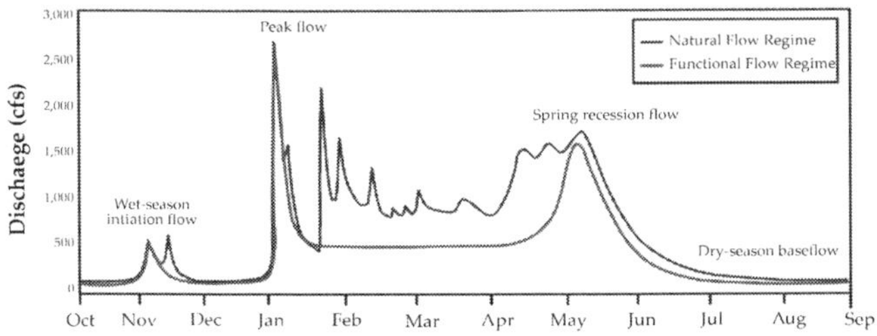

Figure 4.2: Theoretical flow regimes showing flow components for a wet winter/dry summer temperate river.

Source:*https://californiawaterblog.com/page/13/?app-download=android*

Freshening flows in summer and winter play an important role in maintaining water quality by adding flow variability and disturbance to long periods of stable flow. These flows mix the water column, refresh habitat, control salinity levels, transport nutrients and carbon downstream and reduce the risk of excessive plant growth.

High flows, winter bank-full and overbank flows are important to connect rivers with wetlands and the floodplain. It has been revealed that these events entrain organic matter and nutrients from the floodplain into the channel to build and sustain food webs. The organic matter sources are typically plant material from riparian zones that are buried or stored in the channel and slowly break down. These high flow events occur regularly during winter/spring, when the low water temperatures and sustained high flows manage water quality issues such as hypoxic black-water events and increased plant growth that result from increased organic carbon and nutrients.

Most temperate rivers around the world have experienced high levels of regulation which has disturbed natural relationships between water quality and flow and led to water quality decline. Environmental flows studies have been commissioned by river managers on many temperate rivers in order to determine the flows required to restore more natural flow patterns. These studies set minimum flow recommendations to improve the ecological health of rivers. They specify the duration, timing, magnitude, frequency and rate of change for various flow components based on the natural flow regime and the ecological (i.e. fish, macro-invertebrates, vegetation) and geomorphological requirements.

Although water quality is considered in most studies, it is not a major focus when setting specific flow recommendations. The preference has been to recommend complementary works to improve water quality by focusing on reducing catchment contaminant loads at the source through on-ground catchment management actions (e.g. fencing, farm management plans, reducing industrial and sewage treatment plant discharges). Restoring a more natural flow regime is likely to have positive water quality outcomes through improving ecological structure and function. However, through developing a good understanding of the catchment-specific relationships between water quality and flow (e.g. through modelling/monitoring), it may be possible to include more water quality focussed environmental flow recommendations in future studies.

Tropical Rivers

Tropical rivers have a predictable annual hydrological cycle of distinct wet and dry seasons that underpins the seasonal inundation of floodplains during the wet season and the contraction of the rivers to their channel during the dry season. Most tropical rivers are also associated by flooding events throughout the year, but there events are usually predictable than in the case of temperate rivers.

Figure 4.3: Two case scenarios of Tropical River during summer and winter seasons.

Source:*http://www.geocoops.com/landforms.html*

Tropical rivers in most places around the world can be characterised based on tropical wet-dry climate and tropical monsoonal climates. Tropical monsoonal rivers usually have a more evenly distributed rainfall pattern than most tropical wet-dry-rivers. For instance, most stream-flows usually stop during the dry seasons for the tropical wet-dry rivers. However, in some rivers, the flow in the main river channels is usually sustained perennially by base-flows sourced from groundwater and from additional inputs from a more evenly distributed rainfall pattern in the monsoonal climate of head water catchment. One key distinct variation in terms of discharge and the interaction of the river flow with the landscape in tropical rivers result in strong seasonal variation in water quality. For instance, most biological, chemical, and physical alterations usually occur during the different stages of the river's hydrograph. During the dry season, perennial rivers are characterised by base flows and stable hydraulic habitat with low velocities and low suspended particulate matter. Consequently, water clarity is often high. Such physical conditions together with the high light availability in tropical regions are very favourable for in-stream primary production, particularly attached macro-algae.

Generally, the small quantity of nutrients available in most of the tropical rivers, especially those that are typically in areas of low development usually limit the accumulation of various primary producer biomasses. On the other hand, oxygen supply through photosynthesis is a significant element in maintaining dissolved oxygen at concentrations to sustain fish and other invertebrate communities, particularly when accumulation and decomposition of various organic matters during the low flows may lead to high rates of oxygen consumption by bacterial respiration.

Some places, especially along the coastal line have high rate of evaporation. For instance, this may range from 2,000 mm in various monsoonal tropical rivers and about 3,200 mm in various wet-dry-tropical rivers annually. Evaporation is usually greatest at high temperature and low humidity. Additionally, as the river discharge declines or ceases especially during the dry seasons, evaporation rates usually increase because of the longer water residence times. The effect of water loss by evaporation on water quality during the dry season can result in significant increases of dissolved and suspended fractions including nutrients, particulate matter and salts. Electrical conductivity (EC) can be an effective indicator of increased water loss by evaporation because the concentrations of major ions are not affected by the biophysical processes that influence nutrients and other dissolved or suspended fractions, and consequently they increase in accordance with predicted evapo-concentration rates. However, it may not be a good indicator in groundwater-fed systems if the groundwater is saline.

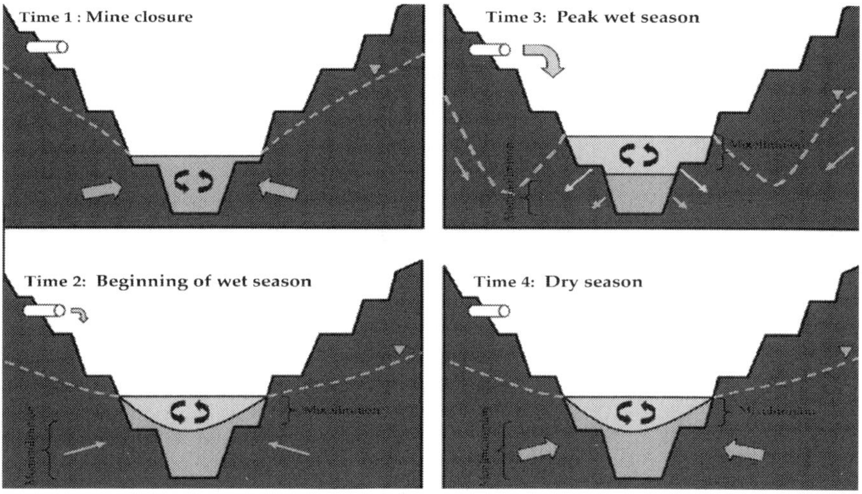

Figure 4.4: Conceptual models for tropical rivers during dry season and flood flows.

Source:*https://www.researchgate.net/figure/248360259_fig6_Fig-6-Time-series-conceptual-limnologic-model-of-pit-lakes-filling-with-water-from-two*

Studies have shown that start of storm event, especially during the wet seasons usually cause low river flow volumes because much of the rainwater is usually socked into dry soils in the catchment areas. As the soil become more saturated and storm activity increases, catchment run off usually result in distinct river discharge events. The initial river flush are usually very rapid and may significantly change the quality of the water over a short period of time. This is because of the interaction between water quantity and the rainfall run-off with the terrestrial environment. The quality of water in the initial run-off is usually acidic (pH value ranging from 5.8-6.0), turbid and with very high concentrations of both organic and inorganic particulate matter and other dissolved nutrients.

A substantial proportion of the mobilisation and downstream transportation of the nutrients and suspended sediments during the wet season occurs during the rising limb of the hydrograph of the first high flow events and may last only short periods (several days) with concentrations fluctuating rapidly. The amount of particulate matter in suspension during the rising stages of early flow increases with stream order and concentrations may exceed extremely high levels (10,000 mg L-1) for brief durations in large river catchments with erosion-prone soils.

The rapid change in water quality and quantity during and after first flush events has strong effects on biological processes. The increased depth and turbidity together with often overcast conditions reduce light availably in the water column and river bed. The increase in velocity exerts drag disturbance on attached algae and macrophytes and reduces biomass. As a consequence primary production and the supply of oxygen are substantially reduced. Oxygen demand is increased by the import of organic matter into the river, which drives high rates of respiration from microbial decomposition. When the first high flow events result in very low concentrations of dissolved oxygen, the hypoxic conditions can be hazardous for aquatic fauna and have been identified as causes of fish kills.

The longitudinal transport of dissolved and suspended fractions during floods is an important process both in the river itself and in the receiving coastal waters. Due to the magnitude of the wet season flows, the total mass of suspended and dissolved matter is very high and essential for the delivery of sediments, inorganic nutrients and organic matter to downstream environments and in driving estuarine processes.

Dry-land Rivers

Most of these types of rivers are found in arid and semi-arid regions and are the most common types of river in most parts of countries. The flow regimes of dry-land rivers are usually unpredictable and may vary from place to place around the world. Most are temporary (unless their headwaters are in tropical regions or are supported by groundwater). This means that most dry-land rivers usually lack a constant surface water flow throughout the whole year. However, they usually experience large floods, followed by extended cease to flow periods where water retracts to a series of isolated waterholes.

Despite their variable water flow regimes, most of the dry-land rivers usually play a crucial role as key water source in most places around the world (for instance, they support significant aquatic and terrestrial biodiversity, including many unique species). Dry-land rivers also play a significant role in supporting both recreational and consumptive water users. For instance, in most places these rivers are used for agricultural activities and to supply water for domestic uses both in rural areas and big urban centres. Dry-land rivers are also used for cultural activities especially by various indigenous communities around the world.

Most studies conducted around the world describe these types of rivers to be having what is referred to as "bust and boom" ecology, which is driven by river flow. The boom phase occurs during large floods and floodplain inundations that encourage high levels of primary and secondary production. This is followed by the retraction of water to refuge pools, known as the bust phase. The permanence of waterholes is usually determined by the evaporative water loss and morphology of the water holes. Some waterholes in these rivers may even persist for more than two years without connection with surface water flow, while others usually dry out quickly. Waterholes can be sustained by surface flows and potentially groundwater contributions.

Along with the hydrology of the river, the quality of water in dry-land Rivers is usually one of the key drivers of changes in the compositional and functional processes of the river. Water quality during the "boom" phase is usually driven by the large volumes of catchment runoff. Flooding also entrains most of the nutrients and organic carbon from the productive floodplain areas into the channels of the river in order to support food web.

Under the "burst" phase, when the river waters retract to the waterholes, water quality is usually driven by various processes such as compound precipitation and concentration, influence of groundwater, and evaporation. Water quality conditions can be harsh at the local scale from low dissolved oxygen levels, high temperatures, increasing salinities, hardness, alkalinity, and cations. In-situ primary production by benthic algae usually acts a key source of energy that plays an important role in supporting and sustaining food web in various waterholes. It provides significant food source for different organisms including turtles, fishes, and other vertebrates living in these waterholes. Sustainable high turbidity levels limit light availability in many of the dry-land river systems. Most microorganisms such as benthic algae can only grow where the bottom of the water hole is shallow enough in order to facilitate at least 1% of all incoming sunlight. Consequently, studies have shown that turbidity of water and the interaction between morphology of waterholes and water volumes influence significantly the production of the microorganism in those waterholes. This usually leads to the 'bath-tub hypothesis', whereby highly productive bands of benthic filamentous algae grow around the perimeter of the waterhole in the photic zone.

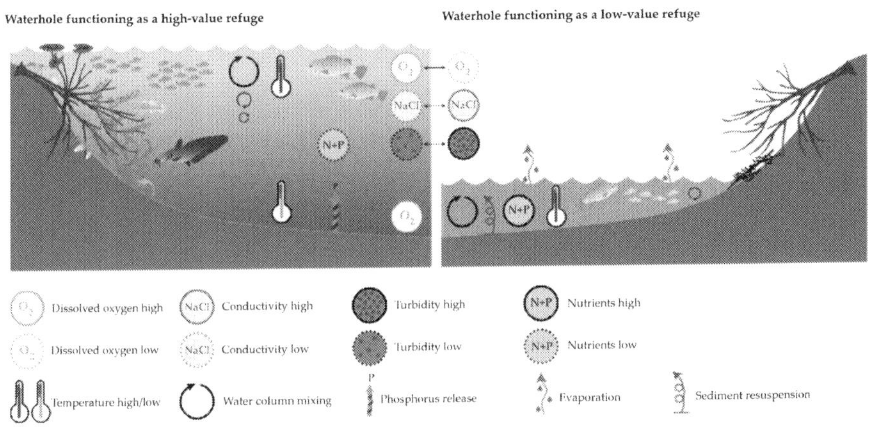

Figure 4.5: Conceptual model of a dry-land river isolated pool.

Source:*http://www.finterest.com.au/fish-habitats/the-importance-of-water-holes-on-our-northern-murray-darling-basin/*

Water quality naturally changes temporally and spatially in dry-land rivers. This makes developing and applying water quality guidelines and trigger levels very difficult for these rivers. One possible solution is to develop guidelines for both the no-flow and flowing phases.

MODIFIED RIVERS

There have been extensive modifications on the regimes of the rivers by human beings around the world. Most river regulations have also been introduced in order to efficiently control and manage movement in catchment areas. These regulations ensure that there are adequate water supplies for domestic use, industrial use, and agricultural use in most places, especially in developing countries. Due to increase in world population and urbanisation in most places, ecological stream degradation has been on rise in a manner commonly described as "urban stream syndrome". Salinity intrusion into stream by groundwater, especially in agricultural landscapes has become a huge water quality issue. This section of the chapter is going to discuss some of the key modified rivers and various major issues on water quality (such as salinization, urban pollution, and cold-water pollution).

Regulated Rivers

River regulation relates to measures taken to efficiently control water movement in a river system for economic gain. It supports beneficial uses such as potable water supply, irrigation and hydro power generation. It can also support social values such as navigation, recreation and flood mitigation. Dams and/or weirs and the associated head-works and channels are built to harvest and control the flow of water. River regulation then determines the frequency, the magnitude and manner of flow releases downstream.

Figure 4.6: Illustrating a regulated river.

Source:*http://www.revelstokemountaineer.com/bc-hydro-submits-application-for-sixth-generating-unit-at-revelstoke-dam-17813*

In addition to the impacts of construction and inundation of habitat through the building of dams, weirs and infrastructure, the operation of dams through river regulation also affect aquatic ecosystem values in the river catchment. This can occur through seasonal reversal in the flow regime (particularly in irrigation supply systems) and thermal pollution from cold-water releases. Dam constructed on the downstream may be affected differently by the river regulations based on the river flow regime. This is because it may either cause more variation or more constant flows and/or increased or reduced flow volumes. For instance, a dam that is constructed downstream for irrigation purposes may cause short-term variation on the river flows especially during the constant flows and during peak demand. Long-term flows reduction may also arise through evaporation and diversions of the river.

There are environmental studies that have been done on the river reaches downstream of key dams in order to determine the minimum flows needed to restore the natural river flows regime to maintain and/or improve the conditions of ecological values. However, aquatic organisms have water quality requirement as well as flow-habitat requirements. One of the major water quality issues is the thermal pollution which is associated with water that is released from storages.

Thermal pollution commonly refers to a significant reduction or increase in the temperature of water compared to the natural conditions such as water cooling through power station discharge and dam release. During the warm months of the year, most large dams' discharges are usually colder and vice versa. This is because water in these dams usually cool and heat at a slower rate due to large volumes of water compared to the shallow, flowing rivers. Cold-water pollution from dam releases is an important issue in most parts of the world because of the varying hot and dry climates. It is usually caused by unseasonably cold that is released from deep layers of thermally stratified dams especially during the summer seasons. The following are some of the key aspects of cold water pollution;

- Dams are typically most strongly stratified in summer and mixed in winter
- Summer depressions in temperature are typically around 8–12° C downstream of major storages that release from below the thermocline, although may be up to 16° C compared to natural conditions
- Winter elevations in temperature downstream of reservoirs are typically 2–4° C, although can be up to 9° C
- Seasonal variations in water temperature can be reduced by between 4° C and 10° C. Seasonal variations can also be

delayed by up to 13 weeks

- The length of river affected downstream of a dam depends on discharge rate, river flowing depth and local climate, but effects have been observed hundreds of kilometres downstream on dams of major rivers

Thermal pollution is a key issue that usually threaten the lives of many aquatic fish species and other organisms. The primary productivity of these organisms can be affected greatly by the temperature of the water. Other effects of thermal pollution include change in the decomposition rates of organic matter, saturation of oxygen in the water, toxicity variation of various contaminants, disease spread and prevalence, breeding cycle interference of different aquatic organisms. Thermal pollution has also severe effects on other beneficial uses of waterways, including recreational activities. However, thermal pollution can be controlled and well managed through creation of special structures (such as multi-level off-takes) which permit discharge of surface water. Alternatively, it can be controlled through the use of special equipment that helps in preventing development of thermal stratification.

Other water quality issues associated with river regulation are caused by low variability in flow and overall lower flow magnitudes in some reaches downstream. Low flows are associated with low oxygen content, temperature extremes, increased concentrations of contaminants, eutrophication and salinization. Dams also reduce connectivity along the river length, which has implications for nutrient and sediment transport which can affect downstream trophic structure and function.

Clear-water erosion can also occur downstream of the reservoir. The quality of water downstream can be affected by dam's location along the river, the surface area and depth of the dam, the climate of the region, various operations done on the dam, morphology and the flow of water, and the hydraulic residence time.

Urban Streams

According to Köppen Climate Classification urban streams can be defined as "streams where a considerable part of the contributing catchment that comprises of development where the combined area of the roofs, roads, and other paved surfaces result in an impervious surface area that is characterised by more than 10% of the catchment". Currently, urbanisation is rapidly occurring in most parts of the world from population growth and the mass movement of populations from rural to urban centres. It is important that this continual urban growth is incisively managed. This can be achieved through taking innovative strategies that should ensure that these urban centres and other surrounding environments are protected and managed properly.

Figure 4.7: An example of urban river.

Source:*http://www.greenhub.cat/mediambient_kr/*

Urban streams are valued greatly as environmental, aesthetic and recreational assets. They may support floodplains and riparian zones that become local biodiversity hotspots or refuges, providing corridors for otherwise isolated flora and fauna populations and their habitat. They may also provide benefits such as wastewater disposal, supplying water to the inhabitant, and mitigation of floods. Urban streams are often heavily degraded due to the increased rate and volume and movement of pollutants into waterways by runoff from impervious surfaces. This causes deterioration in water quality, degradation of stream habitats and flash flooding. The consistently observed degradation of the urban streams is commonly described as "urban stream syndrome". Some the commonly seen symptoms of urban stream syndrome usually fall under four categories, that is biodiversity and ecology, stream hydrology, water quality and geomorphology of the stream. Some of the key water quality and hydrological impacts of urban stream syndrome are discussed below.

Stream hydrology-urbanisation usually plays a key role in alteration of stream hydrology. This usually results in runoff occurring more readily and quickly during rainfall events. There are three major pathways for the hydrological impacts. The first involves vegetation removal and this usually results in reduction of evapotranspiration, catchment storage, and roughness of the surface. The second impact involves surface impervious increment which results in reduction water loss through infiltration. The third impact entails direct stormwater and drainage connections resulting in increased hydraulic conveyance efficiency. Some of the hydrological changes that urban catchment usually exhibit include;

- Decreased base flow- the increased impervious areas mean that there is reduced water infiltration. This therefore reduces discharge and groundwater storage
- Small to moderate increases in flow during small rain events. These flows are frequent in urban area and are usually as a result of direct runoffs on the surface in case of small rain events
- Large increases in peak flows from larger rain events, but of shorter duration than natural. That is, stream flow rises more rapidly during storms and recedes more rapidly after storms, which is described as a "flashy" stream flow.
- First flush is an important and distinctive phenomenon which produces higher pollutant concentrations early in the runoff and a concentration peak preceding the peak flow

Water quality-urban stream water quality is highly variable and is usually an important determinant of the overall condition of the stream flow. Changes to the hydrologic regimes of urban areas can affect water quality as the increased runoff velocities lead to erosion and the entrainment and transport of pollutants present on the catchment surface. Generally, surface runoff, commonly referred to as stormwater, usually carry with it different materials (including of both biological and chemical origin) to various nearby receiving water bodies. Most of these materials are known to cause toxicity to both human beings and aquatic organism and may even alter various processes of ecosystem such as nutrient cycling. Some of the key pollutants in the urban runoffs and their impacts include;

- Litter—(paper, plastics, glass, cigarette butts etc.). Litter impacts on the visual amenity of the waterways, as it tends to float on the surface. It can also degrade habitat for aquatic biota and clog the drainage system impeding the flow of stormwater. Some litter contains volatile solids that may be toxic to some organisms.
- Sediments and suspended solids—Suspended solids are transported from streets and paved areas, rooftops, construction sites and other areas. Loads are generally 10–100 times greater in urban runoff than from undisturbed land and can be transported and deposited at any time as flow velocities decrease. There are both physical and chemical impacts associated with sediment transport. The physical impacts include siltation and smothering of ecosystems, blocking sunlight and reducing water clarity. Chemical impacts relate to the transport of pollutants, such as nutrients and pathogens, which attach to the sediments.
- Nutrients- most urban streams and runoffs are usually loaded with different types of nutrients. This causes eutrophication and aquatic growth stimulated by nutrients entering the water, which can alter the visual appearance of the water, lowering its beneficial value. Visual impacts may include floating matter (algal blooms) and slimes. Other impacts can include dissolved oxygen depletion and objectionable taste and odour from algal blooms.
- Heavy metal- stormwater runoffs from urban areas (including roads, roofs, and industrial sites) usually contain significant loads of different types of heavy metals. Most of these heavy metals are usually a major concern on the health of human beings because of their high toxicity when consumed.
- Hydrocarbons-hydrocarbons can enter into stormwater

from vehicles wear, emissions and chemical spills. They act as potential toxins to both human being and other aquatic organism. Studies have also shown that hydrocarbons pose a threat to the food chain. Hydrocarbons can also cause oil slicks on the surface of water degrading the visual amenity of waterway.

- Organic carbons- these substances act as key pollutants in most of the urban runoffs. Studies have shown that organic carbons usually reduce amount of oxygen that dissolve in water through microbial consumption. This usually results to anaerobic conditions that result to the death of fish, slime growth, water discolouration, and foul smell on water.
- Pathogens—Pathogens are sourced from animal faeces and sewage overflows. They are vectors of disease which can reduce the recreational and aesthetic amenity of the receiving water body.
- Temperature- most of the stormwater runoffs in urban usually have elevated temperatures, mainly because most of these runoffs usually pass through impervious surfaces that is heated. Warmer water can seriously increase the growth of algae through physiological stimulation processes. Other stream adapted to cooler temperatures may suffer thermal stress too.

Groundwater-Fed Streams

One of the key factors that determine the water quality in surface systems is the extent of surface water and groundwater interaction. In other words, streams can be gaining or losing. For a gaining stream, the water flows most of the time into the stream, and its quality is largely or partially a function of the quality of groundwater. Groundwater is very essential in sustaining the flow and/or refuge pools in most of the rivers, especially during the dry weather periods. On the other hand a losing stream is one that water leaks most of the time out of the main watercourse and is usually recharged back into the groundwater system. A losing stream most of the time will create a beneath groundwater zone that is fresh.

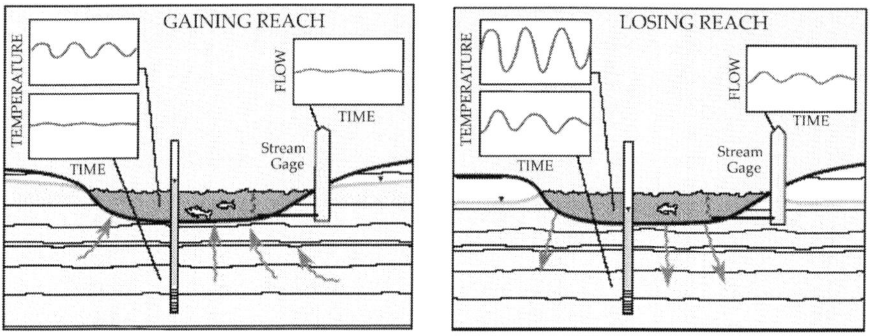

Figure 4.8: Illustrating gaining and losing groundwater stream.

Source:*https://pubs.usgs.gov/fs/2004/3010/*

Most of the water-table of groundwater stream around the world has risen due to native vegetable clearing and over exploitation of the stream water for agricultural uses. This has caused most of the water to become saline.

Saline groundwater results from the long term action of rainfall bringing ocean salt inland. Water quality affected by this mechanism is known as 'primary salinity'. Saline water bodies from groundwater interactions are present in most parts of the world's water bodies, Human-induced salinity is known as 'secondary salinity'. Large-scale irrigation developments are a major cause of secondary salinity. The addition of irrigation water as an input to the hydrologic cycle causes increased runoff and increased recharge to the groundwater system, bringing the water table towards the surface. The added irrigation water can often amount to some 400 to 800 mm/year in areas typically having 250 to 450 mm of rainfall. Consequently the emergence of salinity within a decade or two of irrigation commencing is not uncommon.

Dry-land salinity can also occur where clearing of deep-rooted native vegetation decreases evapotranspiration on the output side of the hydrologic cycle. This too causes increased runoff and increased recharge to groundwater. However, it has been pointed out that climate also plays a role in dry-land salinity. Salinity can occur in ephemeral streams in catchments as small as a few square kilometres, through to major regulated rivers with catchments of around one million square kilometres. The effects of salinity on stream ecology can be severe. Saline groundwater intrusion into streams causes salinity levels to rise. Increasing salinities can exceed the salt-tolerances of many human being, fish and invertebrate species. Some freshwater algae and aquatic plants also have a low tolerance of salt. Therefore, increasing salinities may cause a shift in aquatic fauna and plant communities towards one that is more salt tolerant. Hence, some sensitive native species are lost from the system.

When the groundwater base-flow salinity is relatively high, and the stream flow salinity is low, saline stratification can occur. In such cases, a fresh surface layer floats upon a saline bottom layer. There is limited mixing between the two layers, which results in high salinity and low dissolved oxygen levels in the bottom layer. This reduces habitat for fish and other aquatic organisms. The surface layer can experience extremes in water temperature and can create still, calm conditions suitable for the growth of algae in slow or non-flowing pools.

Groundwater intrusion into riverbeds is considered a very likely dominating process in the potential formation of acid sulphate soils (ASS) in inland river and wetland systems. In affected areas, the groundwater creates a saturated environment within the base of the river during extreme low flow periods (prolonged anaerobic conditions) providing a source of dissolved sulphate (SO_4^{2-}) and iron minerals that creates an environment very conducive to ASS formation. Dissolved sulphate, organic matter and dissolved iron minerals must be present within an anaerobic environment long enough to cause reducing conditions and allow sulphate reducing bacteria to convert sulphate to sulphide. Any free sulphide that is produced can bind to free iron and accumulate in the sediments.

Freshwater inflows are important to control the salinity levels in saline-affected streams. In the first flush after a rainfall event, the initial front of surface runoff displaces the saline dregs in the stream, and smaller sub-streams wash salt accumulated on the surface into the main stem. The result is a spike of salinity on the leading edge of the runoff event.

Typically salinity concentrations are lowest during flooding flows because the flow is high enough to dilute the salt load. The pressure of the surface flows can reduce groundwater intrusion into the waterway. When high flows recede, the water that has infiltrated the banks of the river/creek (bank storage) returns to the creeks and groundwater base-flow continues. The groundwater base-flow tends to dominate so this phase is characterised by increasing salinities as the salt load and flow dwindle away.

Low flow/cease to flow conditions has higher salinity from a high proportion of intruding groundwater compared to fresh surface flows. Evaporation from the water surface in the rivers/creeks can also concentrate salinity levels.

RESERVOIRS AND WETLANDS

This section describes the relationship between water quality and quantity in three types of non-flowing (lentic) water bodies. These are the wetlands, lakes and reservoirs.

Freshwater Wetlands

Freshwater wetlands are saturated areas that are inundated with water either permanently or seasonally. There are many different types of wetlands around the world. They are frequently classified based on;

- Water depth
- Frequency of inundation
- Salinity
- Dominant vegetation

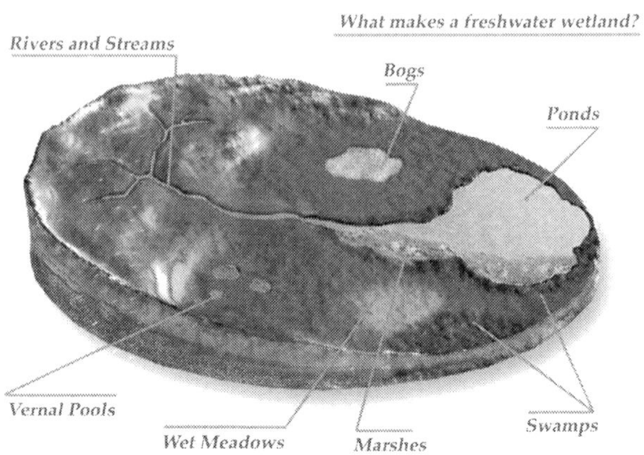

Figure 4.9: Different types of freshwater wetlands.

Source: *http://www.dem.ri.gov/programs/water/wetlands/*

The main classifications of freshwater wetland groups are lacustrine wetlands (e.g. deep, permanent freshwater lakes), palustrine wetlands (e.g. shallow floodwater in swamps) and riverine wetlands (e.g. associated with river channels and floodplains). Wetlands support livelihood of many people, aquatic plant communities, water bird populations and macro-invertebrate communities. Fish populations can also reside in permanently inundated wetlands or use them intermittently for breeding. Wetlands are also valued for water treatment and purification, flood mitigation, recreation, aesthetics and cultural purposes.

The hydrology of wetlands depends on their level of connection with water sources. Some wetlands are connected to rivers. Other wetlands may intersect groundwater. Still others can be isolated and located within low points in the landscape that receive rainfall runoff from the catchment. In urban settings, wetlands can receive water from stormwater drains or industrial discharges.

The wetting and drying cycles of wetlands can be important ecologically as well as for water quality purposes. In natural systems, wetlands typically fill during the wet season and slowly dry during the dry season. The wet season filling refreshes water quality, dilutes accumulating ions and toxins, entrains organic matter for food webs, waters fringing vegetation communities and draws colonially nesting water birds to the site. It releases nutrients from the soils and organic matter, which encourages the growth of seeds in the refilled wetlands. The wetting cycle also causes many invading terrestrial plants to die giving habitat for valuable aquatic or semi-aquatic plants. High flow/flood events in nearby rivers may fill river connected wetlands through flood runners or distributary channels during the wet season. This provides movement opportunities for fish into these wetlands, which provides a variety of food sources, breeding habitat and nursery conditions for some native fish species as the wetlands dry.

The slow drawdown of water during the dry season exposes mudflats that provide food for water birds and habitat for aquatic plants. Some aquatic plants that cannot grow in dry conditions remain dormant until the next flood. Other semi-aquatic plants, such as river red gums, rely on the dry season for their seeds to germinate and allow their seedlings to take hold. The wetting and drying cycle is therefore very important for maintaining a diversity of plant species.

The wetting and drying cycle is a driver of ecosystem functions in wetlands that influence water quality outcomes. Wetting and drying cycles are important for releasing carbon and nutrients that promote subsequent growth by algae, bacteria, plants and animals.

Nutrient and carbon cycling rely on the wetting and drying cycle of wetlands. This is an important feature of wetlands that is utilized for water treatment outcomes in urban and rural environments. Wetlands are often used as pollution mitigation for sensitive rivers or coastal environments. Nitrification and denitrification processes that remove nitrogen from aquatic systems require aerobic conditions (i.e. during the drying of wetland sediment) to convert ammonia to nitrates (nitrification), then anaerobic conditions (i.e. during the wetting cycle) suitable for denitrifying bacteria to convert nitrates to nitrogen gas (denitrification), thus removing it from the ecosystem. Nutrient uptake by plants during their growing phase also removes nitrogen and phosphorus from the water.

Carbon is entrained into wetlands from the floodplains or other surrounding environment during the wetting cycle. It can accumulate in the form of leaf litter on the margin of the wetland especially during the dry weather period. Most particulates and other organic carbon dissolved in the water sources are key elements that supply energy into the water source food web. This is usually through organic matter breakdown (respiration) and primary production (photosynthesis). These functions are also essential in enhancing oxygen infiltration into the water sources. Primary production in river channel is mainly influenced by the amount of carbon coming from the floodplain. Therefore, flooding is one crucial event that helps in movement of the carbon from various floodplains wetland into the river water channel.

Water quality issues arise when the filling of wetlands occurs too infrequently (e.g. in regulated systems or during droughts). This causes the accumulation of large quantities of organic carbon-rich matter on the margins and floodplain. When the next flood occurs, all the material is transported into the wetland (and associated rivers system) and overloads its functioning capacity. The result is a black-water event. Black-water event is typically associated with acidic, low dissolved oxygen condition with high tannin concentrations. Tannins are naturally occurring plant polypherols that usually acts as key defence mechanisms in most plants against different herbivores, pathogens, and other hostile environmental conditions. Most often tannin will persist in water until flushed, however if enough sunlight (UV) enters the water the tannins are usually broken down into carbon dioxide. Black-water also adsorbs more sunlight and may even heat up more quickly compared to the clear water. This may cause heat shock to most of highly susceptible aquatic organisms.

Lakes

Lake systems of the world are highly complex drainage structures that occur on the coast and inland regions in most of the continents. There are different types of lakes that are found around the world. These lakes are classified as below;

- Permanent freshwater lakes (covering more than 8 hectares of land) and they also include large oxbow lakes.
- Seasonal/intermittent freshwater lakes (covering more than 8 hectares of land)
- Permanent brackish/saline lakes.
- Seasonal intermittent saline lakes.

Permanent freshwater lakes are also termed as lacustrine wetlands, are typically deep and fed by a reliable water source, except during times of drought. The morphology of these different lake types is primarily determined by the origin of the lake and underlying geology. However, it is climatic conditions and the nature of inflows that determine the water levels of these lakes. *Seasonal/intermittent lakes* may fill during certain climatic conditions such as La Niña years, whilst permanent lakes may dry out during El Niño years.

Lakes can be closed or open systems, with water quality of open systems largely dependent on the quality of inflows, residence times and flushing rate. The diversity of biota is also greatly influenced by inflows and residence times. According to studies the abundance of invertebrates and fish in lake environments fluctuates greatly relative to inflows. Birdlife tend to respond later in the flood cycle and where there are more persistent waters. Where waters are more permanent there is a mosaic of environments that can support a variety of invertebrates and biota that feed on these invertebrates such as birds and fish. Freshwater lakes in the arid zone are typically episodic due to the high variability of flows and the elevated evaporation rates experienced in these regions. They are generally terminal lakes that have relatively high evaporative losses. A number of lake systems around the world are threatened by water abstraction and river regulation. Coongie Lakes on Cooper Creek and the Paroo Lakes in the Northern Murray- Darling Basin in Australia are primary examples of systems that could be influenced by irrigation practices upstream. At the end of the Murray-Darling system, the Lower Lakes are affected by a combination of river regulation and abstraction. Abstraction and regulation of flow can reduce lake's water level variability i.e. the natural flooding and associated fluctuations in water level that trigger a number of ecological responses. Without this variability some habitats may be lost or impaired. For these reasons, variability of inflows should be a key focus for water managers responsible for the health of lake systems. There are two key physical factors that influence the water quality of freshwater lakes. They include wind stress and water level fluctuations.

Wind stress-Surface mixing that arises from wind stress can be observed as waves on the lake surface. However, depending on the energy and dynamics of the wind there may also be vertical mixing attributed to turbulence. As lakes generally have lower velocities relative to their tributaries, wind stress can often be the driving factor of water quality, particularly where a lake is shallow and the turbulence leads to resuspension of materials on the lake bed. Wind driven circulation is common in shallow lake systems and can promote complete mixing.

Wind stress can have a significant effect on thermal stratification where it influences the middle layer of the water column (metalimnion). It has been stated often that where the top layer (epilimnion) is shallow relative to the middle layer and bottom layer of a lake, the surface layer may either be very active and well mixed relative to the layers below. However, wind action is something relatively beyond the control of water managers and other devices are used to promote lake mixing such as aerators.

Water level fluctuations- Water quality of freshwater lakes is largely dependent on the wetting and/or drying phase. The cycle of wetting and drying of lake systems has been examined by many researchers who developed conceptual models for inundation of a number of different lake types such as commonly wet freshwater lakes, periodically inundated freshwater floodplain lakes and non-floodplain (depression) lakes, and saline lakes. The inundation phase was directly linked to water chemistry and character. Discussed below are the key changes in water quality associated with inundation and recession of water level in a freshwater lake.

Inundation—water level rises

Nutrients can enter lake systems attached to particles or in dissolved form when a flood occurs. Sediments on the lake bed can act as a store for these nutrients which accumulate over time and support complex food webs. Nutrients and organic matter are imported into lakes via floodwaters and nutrients are rapidly released through decomposition of organic material. This can lead to high productivity, which, when combined with greater habitat availability associated with a rising water level, supports greater biodiversity. The uptake and transformation of nutrients and carbon, when lake water levels rise, is the role of primary and secondary producers.

Recession—water level recedes

Evaporation and reduced inflows may lead to water quality decline. Poor water quality may be attributed to a decrease in the buffering capacity of the system as water levels recede. Observed water quality changes can include increased concentration of nutrients, turbidity conditions, an increase in salinity and potentially thermal stratification depending on ambient temperatures. For example, a decrease in inflows to lake may correspond with a dramatic increase in the salinity levels of a lake as large flow events that would otherwise flush the system via an effluent stream (Salt Creek) that has not occurred for several years. In extreme cases, permanent freshwater lakes can recede until they dry out. In instances where a fall in water levels exposes lake sediments, there is potential for ASS to be exposed and become oxidized.

Reservoirs

These are artificial or natural occurring lakes used to supply freshwater for domestic, industrial, and agricultural use. Reservoirs can be created along the course of a natural river through construction of large dams. Reservoirs can also be created off-stream through excavation of the land or through the use of other conventional mechanisms. Most reservoirs around the world are usually constructed in catchment areas that receive high-rainfall. They are typically use for harvesting and storage of water, especially during the wet periods in order to facilitate adequate water supply.

Figure 4.10: *Freshwater reservoir constructed along the river course.*

Source: *http://www.conneryscottishwalks.co.uk/megget_reservoir_to_talla_reservoir.html*

There are several elements that influence the quality of water in reservoirs. These elements include the magnitude, frequency, seasonality and duration, inflows quality, and in-lake processes. River water that flows or is pumped into reservoirs usually plays a huge role in transportation of sediments, different nutrients, and other types of contaminants. Most of these materials usually affect the quality of water in reservoirs in one way or another. Reservoirs that harvest water for potable drinking water supplies are best placed in protected forested catchments where rural and urban activities are restricted. This maximizes the water quality of inflows by reducing the risk of contaminants. Bushfire runoff presents the highest risk of contaminant loads in these catchments. Other reservoirs require a lesser quality of water (i.e. for irrigation) and may have increased levels of development in the catchment.

One of the major water quality issues within reservoirs is algal blooms. There are many types of algal blooms, including toxic blue-green algal species (or cyanobacteria). They affect the amenity of the water, making it unsuitable for potable supply, stock watering and recreation. They can also coarse water treatment processes. There are many factors that contribute to the types and abundance of algal growth as outlined in the table below. Increased light, temperature and nutrient concentrations are the primary factors, the former explaining why algal blooms are most often observed in summer and autumn. Other contributing factors towards the development of algal blooms include still, calm waters with decreased mixing (stratification), water depth and residence time of water in the reservoir.

Stratification is the separation of water column into density-based layers caused by differences in temperature or salinity. Thermal stratification of layers affects deep reservoirs and hence is a more common phenomenon in temperate reservoirs than tropical reservoirs because of the pronounced seasonal variations in temperature. Stratification occurs in the warmer, drier months, when reduced inflows result in minimal mixing within the reservoir. The surface layer of the water column heats up resulting in decreasing density. It then floats upon a colder, denser layer and there is a resistance to vertical mixing. The warm surface layer is known as the epilimnion and the colder water trapped beneath is the hypolimnion. Between the layers is a shallow zone called the metalimnion or thermocline, in which temperature changes more rapidly with depth than in other layers. Once the reservoir has stratified, a large amount of energy is required to breakdown the layers during summer. The breakdown of stratification, commonly referred to as 'turnover' occurs naturally by a decrease in surface temperatures and wind induced mixing. This often occurs in autumn and remains throughout winter and some of spring, until a rise in ambient temperatures may initiate next season's stratification

Table 4.1: Risk factors that can contribute to the risk of poor water quality and algal blooms

Risk factor	**Description**
Average and maximum depth (m)	Shallow water bodies are at greater risk of algal blooms because light is able to penetrate the full depth of the water column and heat to a greater degree than in a deeper reservoir.
Retention time (mean annual lake volume / mean annual outflow)	Short retention times do not allow for nutrients to be recycled and bound to sediments, thus they are available for algal growth. Long retention times usually lead to nutrients removal from the water column through sedimentation and adsorption.
Volume and area quotients (catchment area/lake volume)	The larger the volume or area quotient the greater the influence of the catchment with respect to nutrient loadings and the less in reservoir dilution that can occur.
Strength of stratification	Under stratified conditions nutrients can be released from sediment and later become available for algal growth. The stronger the stratification, the greater the potential for nutrient release from sediments and the greater the risk of algal blooms.
Frequency of algal blooms	Past history of algal blooms provides some indication of the likelihood of future algal blooms. A reservoir with a history of frequent blooms is more likely to bloom in the future compared to one with a low frequency.
Maximum algal abundance	The maximum abundance of past blooms usually acts as potential indicator for the future blooms and reflects the available nutrient status.
Reservoir nutrient concentration	The water column nutrient concentration or annual load usually acts as an indicator for the potential nutrient pool available for algal growth. The greater this pool, the higher the risk of algal blooms.
Nutrient concentration in inflows	Nutrient concentrations in inflows provide an indication of the potential nutrient loads to the reservoir and hence the risk that this poses to blooms.

Source:*https://www.cees.iupui.edu/research/algal-toxicology/bloomfactors*

Cycles of stratification and mixing of the water column are known to drive algal growth in nutrient-enriched reservoirs. Stratified conditions create calm, still and warm conditions at the surface that is favorable for the growth of algae. The bottom layer becomes depleted of oxygen and under such conditions releases nutrients from the sediment, which further drives algal blooms at the surface. Other water quality issues can occur from stratification in reservoirs. This occurs when the reservoir off-take is below the thermocline. Under such conditions, cold-water pollution can occur downstream. Also, the anoxic conditions in the hypolimnion produce a reducing environment whereby manganese, iron, phosphorus, sulfides and ammonia are released from the sediments. These chemicals can alter the natural taste of the water. Additionally, it can also stain the water, which makes it displeasing for potable water supply customers.

CHAPTER 5

WATER POLLUTION

INTRODUCTION

According to the Draft Environmental Management Act, water pollution can be described as "the direct or indirect introduction of foreign substances (as a result of human activities) in to clean and safe water which may be harmful to human beings health, well-being, or may interfere with amenities and other legitimate uses of water". In this chapter, water pollution is described according to the medium into which it is introduced. Thus the chapter has been broken up into several headings and sub-headings. Water pollution in most parts of the world covers any pollution which enters both inland and marine water bodies. These include oceans, ephemeral rivers, shared rivers, and underground aquifers. Because of the close association between water and soil pollution, soil pollution is also discussed slightly in some parts of the chapter. Marine pollution describes any pollution that enters the coastal waters either directly (e.g. direct discharge of effluent into the ocean) or indirectly (polluted water from rivers washing into the ocean).

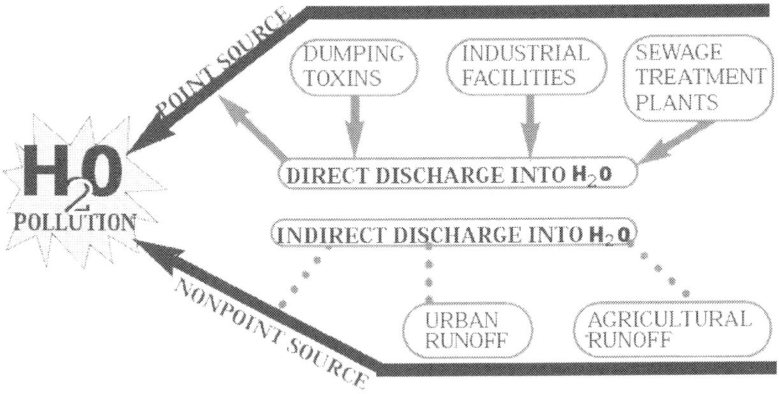

Figure 5.1: Water pollution sources.

Source: *http://cpcbenvis.nic.in/water_pollution_control.html*

SURFACE AND UNDERGROUNDWATER POLLUTION

Pollution of freshwater is the most serious form of pollution in most parts of the world. It is the most widespread and poses the most severe health risks especially in most arid and semi-arid areas. Additionally, very arid countries can ill afford to let any of their potentially drinkable water become unusable. General water pollution is caused by many factors, although almost all of them originate from human activities. The major causes of water pollution are industry and the improper disposal of household waste (i.e. solid waste and sewage).

Faecal pollution- this is probably the major form of water pollution especially in most of the developing countries. Faecal water pollution comes from both human sewage and from animal excrement. Faecal pollution from livestock enters the water in two main ways: first, animals that drink near water sources leave very high concentrations of dung in the surrounding area, this can then seep into the aquifer or enter the source directly during a flood. Secondly, livestock that drink near bodies of surface water may have their dung washed into the water during the next rain, which contributes to nitrate and faecal pollution.

While some animal excrement is a normal and healthy aspect of a natural environment, the high concentrations caused by farming contribute to nitrate and faecal pollution. Run-off within a catchment area will always contribute to dung being washed down the river to the nearest water body.

Although not as large a problem associated with livestock waste, human sewage is also a contributor to water pollution. This is mainly due to inadequate facilities and/or poor management practices. In most of the developing countries and least developed countries, many sewage ponds (i.e. oxidation ponds and evaporation ponds) either have leaks that let sewage seep into the ground or have problems with flooding over when it rains. This dumps sewage and all the associated health risks directly into the surroundings. According to a survey that was conducted in Africa on sewage system by officials of the MET revealed that only few are properly maintained and operating. The majority of sewage systems in most African countries are either broken down or are poorly managed and maintained.

Faecal waste tends to introduce two main types of chemical pollution into water, nitrates and sulphates. Nitrates are dangerous to infants under one year of age. Excess levels of sulphates cause diarrhoea and dehydration. E. coli is also present in faecal waste, and can also lead to diarrhoea and dysentery.

Solid waste disposal sites- Improper location of solid waste disposal sites is also a major contributor to water pollution globally. Water that seeps through a landfill will pick up various impurities, depending on the waste it passes through. A landfill should be placed on an area that will not allow penetration by water that flows through the waste. At the very least, care should be taken to ensure that water seeping through the landfill never enters any sources of drinking water. Ideally, hazardous waste materials should be separated from other waste and placed in a special lined cell to prevent water from passing through. The municipals and other privately owned utility companies should monitor the landfill sites by analysing water taken from strategic sites around the country. Such initiatives should be taken by both local and national government in order to ensure that the well-being of people is properly achieved. This is due to a combination of lack of awareness among the persons responsible, and lack of money to pay for such measures. Similar pollution problems are caused when individuals dump their waste in riverbeds or bury it near a water source.

Pesticides and other agro-chemicals- Irrigation of fields can cause water pollution. This mainly occurs when farmers mix agro-chemicals (e.g. pesticides, fertiliser) with irrigation water. Generally people do not really know what happens to this water after it passes through their crops:

- It may evaporate leaving the chemicals in the field (ideally)
- It may penetrate through the soil or
- It may flow off the fields into nearby water courses.

Pesticides are a serious source of water pollution even when they are not put directly into irrigation water. In some of the developing countries such as Namibia people are fond of mixing DDT with diesel fuel which is then sprayed onto the houses in order to kill various insects and other types of vectors. It has also been reported that people mix small amounts of DDT with the millet and then spray the mixture to the grain storage baskets. This is done to keep rats/vermin from destroying the stored crops. DDT is a residual poison, and will accumulate in people's and animals bodies (in their fat tissues) eventually reaching harmful or possibly fatal levels. It has also been reported that people use old DDT containers to store and carry drinking water. A study conducted by MET many years ago revealed that DDT affects the eggshell thickness of birds in the malaria/DDT affected areas.

Other miscellaneous pollutants- Fluorides are almost in all cases of natural origin and it is occasionally in other isolated spots of a country. Most countries lack the required technologies for the treatment of water to remove unwanted concentrations of fluoride. In some cases it has been reported that tar from paving roads has been partially suspended in rainwater causing some pollution of water bodies receiving the storm water. Tar contains many toxic materials and carcinogens. This problem is made worse when the unused tar is simply discarded at the side of the road.

WATER POLLUTION FROM SPECIFIC LOCAL SOURCES

There are some sources of water pollution which are particularly intense in specific area in which they are located, but are not found throughout in most parts of a country. One good example among these is mines. Mines often use chemicals to extract the desired metal from the ore. The chemicals used depend on the type of process employed by the miners (for instance, cyanide and sulphuric acid for extraction, and xanthate for flotation). These chemicals are poisonous if ingested, which can easily occur by drinking water that is contaminated with them. Such chemicals are often found in the tailings or effluent from the mine. Just as with a landfill, improper placement of a tailings dam can cause serious pollution problems, both for the surface water and groundwater. This can be a severe problem where the substrate allows easy penetration of water into an aquifer, especially where the bedrock is limestone or dolomite. These rocks are easily dissolved by acidic substances which then allow contaminants to find their way into the underground water, where it can spread far. The problem is exacerbated where the bedrock is fractured or faulted.

Figure 5.2: Lithium mining as a major water pollutant.

Source:*http://www.dailymail.co.uk/home/moslive/article-1166387/In-search-Lithium-The-battle-3rd-element.html*

Petrol stations have a high potential to pollute the water too. This is caused mainly by leaking buried petrol storage tanks. Some aquifers are today still too polluted for use because of petrol contamination that occurred 25 up to 40 years ago. Another potential risk associated with petrol stations/garages is the possibility that they may dump their waste oil directly into the surroundings instead of disposing of it properly. Abattoirs generate biological waste that can cause problems if not properly disposed. Organic waste such as blood, offal and other unwanted animal products should be buried under at least 500cm of soil. For most abattoirs which only slaughter animals, send the skins to a tannery and the meat to a processing plant, this is sufficient treatment. However, some abattoirs have vertically integrated operations which may include facilities for breeding and slaughtering animals, processing meat and tanning of hides. Such operations with varied processes have the potential to produce varied effluents, and different management and control measures need to be implemented.

Tanneries in particular use potentially harmful chemicals (e.g. chromium, sulphide salts, salt for curing) in their tanning processes. If not carefully monitored and contained, these chemicals can cause significant water pollution. Some tanneries operate their own waste water treatment plants. To date none of these plants are operated and managed correctly in most parts of the world, especially in developing countries.

ASSOCIATION BETWEEN WATER POLLUTION AND SOIL POLLUTION

Soil pollution is generally associated with a similar type of pollution in water. When polluted water leaches through soil, it leaves behind some of its polluting substances, which can contaminate the soil and make it less suitable to sustain life. Therefore, any of the above sources of pollution can contaminate soil. Leaking petrol storage tanks are especially harmful to the soil.

One type of water related pollution that is nonetheless specific to soil is irrigation. If the water used for irrigation contains salts and evaporates from the fields, the salts remain behind. Over time this leads to the accumulation of high levels of salts in the soil, even if the water did not originally have very high salt concentration. Highly saline soils are unsuitable for most types of plants, so the pollution renders the soil useless for cultivation of crops.

As mentioned earlier, soil pollution generally occurs in conjunction with water pollution. However, there is one report of an isolated instance of soil pollution that was totally unrelated to groundwater that occurred in Namibia where several years ago a number of full pesticide containers were being unloaded at the Agricultural Depot in Keetmanshoop. During this unloading process some of the pesticide leaked out into the soil. In order to repair the damage that was caused by this accident, the Ministry of Agriculture had to rinse all the nearby buildings and remove 30cm of topsoil in the affected area. This was disposed of at the sewage works and buried together with raw sewage to allow for bacterial breakdown.

MANAGEMENT PRACTICES TOWARDS WATER POLLUTION CONTROL

In most parts of the world, the issues of water pollutions are not adequately managed and controlled despite the impacts received by different people. This is mainly due to inadequate resources especially towards the management and inadequate awareness among the polluters, local authorities, and the public in general.

However, there has been a shift of emphasis in the management of water pollution in most countries. The emphasis is to be on enforcement, whereby the utility companies should try to monitor the activities of businesses, industry and local authorities. When they find practices that are in contravention with the law they should try to force the institution to change its practices or shut down. Unfortunately there have been a number of problems with this approach. First of all it is difficult and costly to effectively monitor the actions of many different institutions countrywide, so presumably many illegal activities are not detected. Second, even when a polluter is caught enforcement is not always easy. Sometimes the polluter may contest the charges in court, when this has happens, the utility company may end up losing the case.

Because of these difficulties, a new approach should be taken whereby emphasis should be on a pro-active solution to water pollution. The utility companies in charge should try to make polluters aware of various health effects and other financial repercussions of heavy water pollution. It is hoped that if people understand the consequences of their actions, they will behave responsibly. Perhaps, and this idea should be elaborated among staffs of the relevant division, staff should then form a team that visits each region or major town and endeavours to explain to relevant people of the region the importance of effluent control, the terminology used in the questionnaires, etc. Such action should include explanations to be given in English and other major vernacular of a region.

MARINE WATER POLLUTION

Marine water pollution is not a problem in most places of the world, especially in landlocked countries. However, those countries with a coastline usually have different legislations and policies enacted in order to manage and control marine water pollution. Most of these policies and legislations are directed towards various activities conducted on or along coastal water such as fisheries, transportation, mining, and other industrial activities.

Figure 5.3: Illustrating industrial pollution of marine water.

Source:*https://blogs.ntu.edu.sg/hp331-2012-astrea/what-are-they-dying-of/local-threats/pollution/*

Generally, most of marine water pollution comes from specific sources and is discussed in the next section. Biocide paints (used to keep off barnacles, etc.) contain mercury, cadmium, lead, chromium or organotin (TBT) or some combination of these. These chemicals can be released into the water when the paint is scraped or chipped off the hull of a vessel.

Marine pollution from specific local sources may occur in different ways. For instance, fish factories discharge most of their effluents as a part of their fish processing activities. There are several different processes that fish factories use, and each can create a different type of effluent. Thus it is possible for a single plant to create four to six different types of effluent, each of which may need to be treated differently. Temperature and density pollution are sometimes not considered to be pollution because they are not 'foreign' materials being introduced into the environment, rather they are naturally occurring substances that are introduced in a different state or at different concentration. They normally do not have severe consequences because they dissolve quickly and easily. However some forms of life are adapted to certain specific environmental conditions and can be harmed by slight fluctuations in these conditions.

Temperature pollution is a significant change in the temperature of some part of an environment caused by human activities. In most parts of the world, the main source of temperature pollution occurs when leftover ice is dumped into the ocean, creating a pocket of cold, fresh water in the sea. Density water pollution on the other hand is mainly in the form of a discharge of highly saline water into the ocean. Discharge from the salt works located along coastal areas may contain high levels of harmful chemicals (such as sodium chloride, magnesium chloride and calcium chloride). The effect of this discharge is mitigated because the effluent is pumped directly into the surface to facilitate mixing of the highly saline water with the rest of the sea water.

METHODS OF WATER CONTROL (PERMIT SYSTEMS, INSPECTION, OTHER)

One of the best ways of controlling water pollution in a country is through Department of Water Affair. This should be done through issuing of exemption permits to different companies, businesses, and other institutions that would wish to discharge their effluents into water bodies. In other word, these exemption permits enable different institutions and companies to discharge their waste or effluents that are not in compliance with the set standards (they are exempted from compliance).

The process of acquiring an exemption permit is usually rigorous. First, the company that needs to discharge its effluent is required to submit a well-prepared application to the department of water affair. Then the department should respond to various questions that entails specific water treatments (such as bio-filters, oxidation ponds, French drains and septic tanks etc.). Once complete, the department should then prepare a technical report following the information that has been provided, and if possible, a physical visit to the company should be done to inspect the status of the effluent been discharged. However, most physical inspections are usually expensive and laborious from the logistical point of view and so are not always possible.

After the technical report has been prepared, the department of water affair should then evaluate this report with the help of other related ministries (such as ministry of health and ministry of social services). Based on the evaluation conducted, the permit is either denied, or temporally permit is issues (which should not exceeds more than one year), or a more permit of five years is issued.

According to Department of Water Affair viewpoints, the term 'exemption permit' is a curse. This is because some of the applicants for such a permit actually treat their effluent so that when it is discharged, it is of a higher quality than is required by law. Thus, they are not really being exempted from anything. Perhaps a system of two permits is an answer. A discharge permit could be issued to those applicants who have the resources and personnel to ensure that their effluent does comply with the law. Other institutions, such as small towns and villages, which cannot reasonably be expected to comply with the established standards, would be able to apply for an exemption permit to allow them to discharge their effluent as it is. Another weak point in the exemption permit system relates to enforcement. The Department of Water Affair does not have the ability to shut down an operation that does not comply with the regulations.

MAJOR ACTORS INVOLVED IN WATER POLLUTION PREVENTION AND CONTROL

Public Sector

In most countries governmental ministries usually act as the primary actor involved in the prevention and control of water pollution. However, there has been a shift towards decentralising this control by putting monitoring, controlling and enforcement in the hands of local authorities. Under the new approach, the government would issue a single permit to a local authority. The permit would have several sections referring to individual organisations or parties operating within the jurisdictional area that do not dispose of their effluents into the sewage system. The local authority would be responsible for ensuring that all industries within its jurisdiction comply with the relevant regulations. The government in turn is responsible of backstopping these efforts by doing spot checks and investigating any complaints made by the public.

Private Sector

Most private sectors are also key players in preventing and controlling water pollution. However there are some industries that do an excellent job by recycling and treating their own effluent or part thereof.

CONSTRAINS IN WATER POLLUTION PREVENTION AND CONTROL

The actual costs of preventing water pollution are unknown. However the major costs involved in monitoring and regulation of pollution levels arise in remuneration of officials and logistics for them to operate. The costs for physical inspections include vehicle and equipment maintenance and laboratory fees.

Another problem that hinders effective and efficient water pollution control and prevention is lack of information that is relevant and accessible. Most of the people do not know when and where wastes should be dumped and by whom. This is particularly a problem when perpetrators dump large quantities of waste infrequently. If polluters are not caught in the act of dumping it is usually impossible to determine who is responsible for the pollution.

Another problem that occasionally arise is the attitudes of some officials, inside and outside the civil service, involved in waste discharge, compliance monitoring and pollution control. It has been stated that some are usually ‘unwilling, unable and ignorant’ - especially in the rural areas. Water pollution control, effluent treatment, disposal and the potential ecological effects constitute a difficult, complicated and often sophisticated task, and require dedicated staffs to be carried out effectively. Proper training is often lacking on all levels.

Another constraint reported is that most utility companies, especially in developing countries usually do not have their own laboratories in which to conduct tests. Instead they have to tender out their analyses to relevant businesses. In the long run this is costly and inefficient, especially since many tests cannot be performed locally and samples need to be sent to Europe and other developed countries.

RECOMMENDATIONS

There should be more physical inspections of sites that apply for exemption permits, and these inspections should be made more efficient.

However, staffs shortages and poor incentives that would help to attract dedicated people make it difficult for the government to fulfil its obligations with respect to water pollution prevention.

It is hoped that through the broad dissemination of relevant information, many people will have the details they need to effectively do their jobs. Training may be more difficult and costly to provide than information, but in the long run it will be less expensive than to allow a country's water systems to be poisoned.

To address the problem of not knowing when pollution is being dumped into the environment, an official website where people could go to anonymously report instances of pollution should be created. If people, particularly school children, are made aware of this site, then illegal polluters will have greater difficulty hiding their crimes.

A common problem is that many institutions have a great deal of raw or semi-analysed data, but these data are not available for one reason or another. Many older records (which are still relevant and very important) only exist in paper form/hard copy. This makes it time consuming and difficult to gain useful information from these data. For instance, it would be helpful to have all data on a certain place available to be able to track trends in water quality (for drinking water or to discover a degree of slow-moving pollution taking place). Another difficulty is that even when data are computerised, they are often fragmented or inaccessible to those people who need them.

To address this situation, it is important to have a centralised database that contains all available national data regarding water. Then any institution that needs such information can retrieve the relevant data quickly and efficiently.

CHAPTER 6

WATER QUALITY PROTECTION

INTRODUCTION

The quality of water can be described as the biological, physical, and chemical characteristics of water. This is mainly with respect to drinking usability. It is the responsibility of every government and other non-governmental agencies mandated with planning and implementation of various water and sanitation activities to ensure that there is adequate quality water for all citizens in the country. Domestic water for drinking and cooking should be supplied from water sources that are safe and clean. Water quantity on the other hand entails the total amount of water that is discharged from a water source. It can also be described using various flow regime aspects such as duration, frequency, and the timing.

The health and integrity of different people around the world highly depend on both the quality and quantity of water available. In other words, the diversity of human societies in different regions is highly influenced by the absence or presence of quality and quantity of water. Water quality and quantity also act as key determinant of survival for human beings and other animals in the world. The United Nations Human Right Council

(UNHRC) in 2010 implemented some resolutions that ensured human beings have the right to access clean and safe water for drinking and sanitation. Actually, access to clean and safe water is considered as one of the key building blocks for social and economic structures in a country. This is because water is used in most of the sectors to support growth and development. These include irrigation sectors, mining, manufacturing, agriculture, and fishing among other sectors. Over history, water has played a key role in shaping the geographic distribution of human populations and their quality of life and culture.

There are six key environmental values associated with water quality and quantity that influence the health, safety, and public benefits. These values need to be protected and they include the aquatic ecosystems, industrial water, aesthetics and recreation water resources, drinking water sources, spiritual and cultural water values, and various primary water industries (such as general water uses, aquaculture water uses, irrigation water resources, and livestock drinking water).

According to World Health Organisation (WHO), it has been recommended that the *per capita* water services per day to be maintained at 20 litres. Different countries around the world have policies that focus on water supply integration, educating people about hygiene and sanitation services using a platform of gender equality, empowerment of women and community planning, supervision and management of water supply services in order to ensure water security throughout the country.

Community participation and management is known to be important to ensure the sustainability of community water projects. Consideration of consumers' willingness to pay for water also ensures more equitable and sustained use of the improved water supplies.

In most developing countries, where water security has been an issue, provision of enhanced water points can play a crucial role in ensuring sustainable supply of clean and safe water for all. However, researches have shown that these initiatives (such as installed hand-pumps in developing countries) have failed due to one reason or another. For instance, the high rate of failure has been attributed to technological reasons (such as poor construction, high cost of spare parts, inappropriate technology) but also to social and institutional factors. Different projects have also been initiated in different countries to address water security issues through village involvement and ownership, financial capacity building and an appropriate technical approach including proper borehole development and the specification for hand-pumps that can be maintained at a village level.

IDENTIFYING MAJOR WATER RISKS

There are different potential threats related to availability or contaminant sources for drinking water that influence water security. Most of these potential drinking water contaminants can be ranked based on the risks likelihood (that is higher risks, moderate risks, and low risks). On the other hand, most of the potential contaminant sources fail within the following categories: above ground/underground storage tanks, agricultural activities (such as the use of pesticides, fertilizers, and animal waste management), dams/powerhouses/upstream reservoirs, fish hatcheries, forest practices (such as clear-cuts, landslides, and pesticides), industrial and commercial facilities, landfill/dumps/illegal dumping/junk yards, recreation (official and unofficial campgrounds, boats, river recreation, golf courses, and parks), resource extraction (such as mines/gravel pits), road vegetation management, storm sewer outfalls, transportation, urbanization (non-point source runoff), septic systems failures, and wastewater treatment plants.

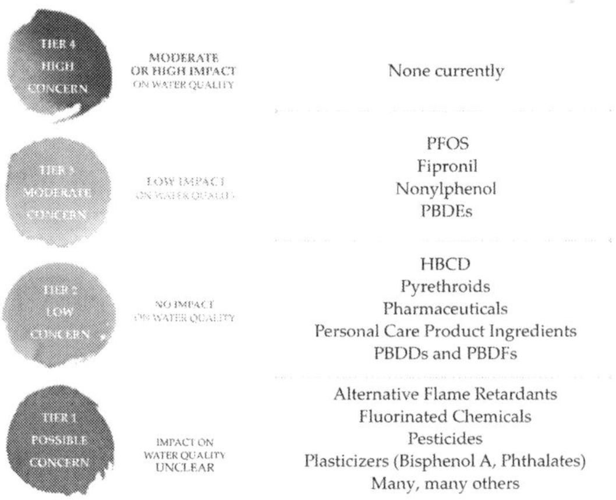

Figure 6.1: Illustrating the risk ranking of different pollutants on water quality.

Source:*http://www.sfestuary.org/estuary-news-old/shifts-on-the-pollution-front/*

Above Ground/Underground Storage Tanks

This is classified as one of the moderate potential risk factor on the quality of drinking water. Inappropriate management of stored items and spills leaks above the ground storage tanks may contribute to contamination of water. Underground storage tanks are also used in most parts of the world and may also represent a threat to the quality of water especially if they are damaged due to improper installation or failure of the pipe connections. Additionally, active underground water storage tank and above ground water storage take may also represent potential source of surface water contamination especially through spills and overfills during water delivery.

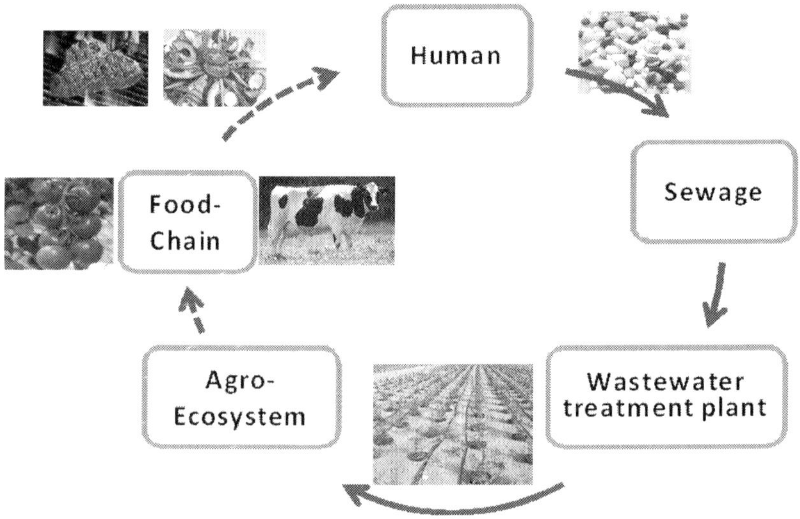

Figure 6.2: Exposure of organic pollutants via irrigation with wastewater effluents.

Source:*http://www.agri.huji.ac.il/utils/single-post-id.php?id=419*

Agriculture

Agriculture is classified as one of the key or higher risk factor of water contamination. Some of the potential non-point source pollutants include sediments, nutrients, pathogens, oxygen-depleting organic and pesticides from irrigation and other non-irrigated plants areas. There are several sources of these pollutants, such as farm animal grazing areas, farm shops for machinery repair, stores, chemical mixing and handling areas, and boarding stales among others. Many different farm chemicals are used on various commercial crops around the world, and they are as key water pollutants among all the other agricultural activities.

Nurseries

These are classified as key or higher risk factors for water security. Different countries practise nurseries crop farming and these usually act as one major source of water pollution. Nurseries contribute to nonpoint source pollution, primarily by nutrients, toxic chemicals, PCP's, antibiotics and medications, and sediment. Container nurseries, in particular, tend to use relatively large amounts of fertilizers, insecticides, and fungicides.

Confined Animal Feed Operations and Animal Management

These are classified as higher risk factors on water quality. There are several permitted Confined Animal Feed Operations (CAFOs) within the different countries. Both permitted and non-permitted CAFO facilities are a concern due to nitrate, bacteria, and other potential pollutants resulting from large numbers of animals faeces in one location. Improper storage and management of animal waste, in addition to pharmaceutical use to keep animals healthy, may also impact drinking water. Concentrated livestock operations may contribute to soil erosion and sedimentation issues.

Dams/Powerhouse/Upstream Reservoirs

They are mainly classified as low-moderate risk factors on water quality. Different countries have constructed large dams on the mainstream of major rivers. Dams and powerhouse operations typically involve the use of fuel, paints, solvents, polychlorinated biphenyls (PCBs) coolants and lubricants. Studies have shown that reservoirs usually play a huge role in contributing prolonged turbidity especially during strong storm events. Turbidity and soil erosion can also be attributed to health watersides, river water fluctuating, and other construction activities along the water source. Water quality may also be affected during transportation of materials, poor chemical handling, leaks, and spillages on the dam.

Recreational uses of reservoirs are also key source of pollution on the dam's water. Stratification of waters in a dam is a source of changes in water quality as in deep reservoirs waters from hipolimnium strata can be liberated downstream; these waters have very low oxygen concentrations related to decomposition processes allowing the liberation of harmful compounds such as methylmercury.

Fish Hatcheries

They are classified as higher risks factor on water quality. There are different fish hatcheries located in different parts of a country. Most fish hatcheries usually discharge effluent that contain nutrients, biochemical oxygen demand (BOD), and other pharmaceuticals (antibiotics). These substances are major potential water pollutants.

Forestry Practices

They are classified as moderate-higher risk factors on water quality. Forestry practices include activities related to growing and harvesting timber. Pollutants typically associated with forestry practices include nutrients, sediments, organics, and increased temperature. Erosion and subsequent sedimentation result from timber harvest, road construction, stream crossings, and high intensity fires. Increased water temperature results from removal of riparian vegetation that shades streams. Tree cutting and yarding is also a potential risk factor on water pollution. This is because it usually increases soil erosion, increased turbidity, and discharge of chemical into the river thus affecting the quality of water. Inappropriate handling and over use of fertilizers and other farm chemicals are also key contributors to water contamination. Private and urban forestry activities should be analysed for their potential impacts too. This is because various activities (such as fertilization, harvesting, and prescribed burning) act as key nitrogen sources which may end up polluting water.

Industrial and Commercial Facilities

These are classified as moderate-higher risk factors on water quality. A variety of businesses including auto service stations, repair shops, dry cleaners, beauty salons, print shops, medical facilities, logging and construction companies, parking lots and retail malls, machines shops, and sand and gravel operations represent other potential nonpoint storm water sources of pollutants. Many of these are located along big towns and cities.

Landfill/Dumps/Illegal Dumping/Junk Yards

They are all classified as higher risk factors to water quality. Most of the illegal dumping yards usually contain hazardous and dangerous substance that can contaminate ground water and surface water. This includes residential accumulations that have the potential to contaminate groundwater with toxins. Poor components and other fluid handling at these dumping sites may also result in to pollution of the storm water runoff pollution. These substances can also infiltrate into the groundwater, thus causing subsequent groundwater pollution. Contamination can also result from stored metals and other hazardous materials because these sites are typically located outdoors where they are exposed to the elements.

Permitted Discharges from Industries

They are classified as moderate-higher risk factors on water quality. Point source discharges on different rivers can be identified primarily from a database of water quality permit holders from Department of Environmental Quality. Facilities with water quality permits, from which permitted discharge and/or runoff could potentially impact the quality of water in different water sources. Bio-solids management and recycled water applications associated with these kinds of permits should also be looked at as potential risk.

Recreations Activities

They are classified as low-moderate risk factors on water quality. Different countries have watersheds used for a variety of different recreational activities. Recreational activities such as boating, floating, hiking, camping, hunting, biking, Off Highway Vehicle (OHV) use, and fishing can contribute nonpoint source pollutants to the river. Potential pollutants associated with outdoor recreation activities include untreated or partially treated human waste, petroleum products, PCPs, detergents, solvents, paints, and non-degradable litter. Overuse of recreational water bodies may increase the risks of erosion on the stream banks thus resulting to turbidity. Boat's emission and fuel spills (especially the older two-cycle engines) may also contribute to hydrocarbon contamination. Emissions refer to hydrocarbon emissions/pollution emitted by gasoline engines. This is mostly an issue associated with the two-cycle engines that are smaller, older and cause more pollution. The larger 4-cycle engines are generally more efficient. Pesticides and fertilizers from golf courses and parks can also lead to degraded river water quality.

Resource Extraction (Mines/Gravel Pits)

They are mainly classified as low risk factors on water quality. Several gravel quarries and mines are found within watershed areas. Gravel mining is associated with increased sedimentation, petroleum hydrocarbons, and metals. Inappropriate chemical waste handling and spill leaks from machineries used in mine sites may also act as key non-point source of water pollution. Gold mining for instance is a high risk for water security including mercury in the form of methyl mercury which is accumulated along the food chain in rivers, wildlife, fishermen and fish consumers.

Roadside Vegetation Management

They are classified as moderate to higher risk factor on water quality. Untreated road runoff finds its way into the river via roadside ditches and streams. Potential pollutants include sediments, petroleum products, metals, and herbicides. Over use and inappropriate pesticides and other farm chemical handling may also impact the quality of water in a huge way. Different governments mandates that state agencies shall implement integrated pest management (IPM) practices when engaged in pest control. Studies have identified roads and roadside maintenance as potential sources of the pesticides previously identified to be present in the river water.

Storm Sewer Outfalls

They are classified as higher risk factor on water quality. Urbanization is characterized by many types of impervious surfaces, including parking lots, building roof tops, highways, sidewalks, and driveways. Sediment and chemicals may accumulate on these surfaces and be washed off into storm drains or ditches during heavy rainfall. Numerous potential sources of contamination, due to runoff and/or accidental spillage/release may be created in any area where surface runoff reaches rivers and other water sources. These point sources have the potential to carry gas, oil, and grease products, as well as heavy metals, pesticides, herbicides, nutrients, and other toxic chemicals and in many but not all areas are regulated by permits. Human and animal pathogens are other likely contaminants.

Transportation Activities

They are classified as higher risk factors on water quality. Several major transportation corridors and bridges are located within river basin watersheds. Truck movement on a country's highways are involved in the shipments of a wide variety of materials, but gasoline and diesel are probably by far the most common products. Road building, maintenance, and use may increase erosion and slope failure causing turbidity. The rapidly increasing rate of vehicle transportation in the world increases the risks of water pollution through fuel spillage and leakages. Transportation of hazardous materials over bridges and the railroad bridges is a particular concern for various utility companies and other municipals of different countries.

Urbanization

It is classified as moderate-higher risk on water quality. Storm water runoffs from urban areas represent both a water quantity and water quality problem. Development and construction are two major contributors of non-point source pollution. Poor storage, use, and disposal of household chemicals, pesticides and fertilizers may seriously affect the quality of river water in a huge way. Additionally, poor handling automotive chemical handling contribute to water pollution through leakages and spillages. The issue of poor housing due to over population in urban centres has potential impact on the quality of water, especially if there is mismanagement of various utility services (such as septic systems). Over watering (irrigation) in urban areas is also a conduit for phosphorus, fertilizers, and pesticides to enter the river. Increased runoff is associated with areas where urbanization has converted open-space land to impervious surfaces. Most developing countries are currently experiencing the heaviest pressure for residential development in urban areas and this has resulted to increase in water pollution. Some of the key pollutants that can be identified from surface water runoff in most of the watershed in urban areas include toxic chemicals, petroleum by-products, heavy metals, different nutrients, and other sediments. Metals (such as copper, lead, nickel, and zinc) have been shown to exceed federal water quality criteria in many urbanized basins in different countries.

Septic System Failures

They are classified as moderate risk factors on water quality. Nonpoint source pollution originating from septic systems is attributed to failing or malfunctioning systems. Sewage disposal systems typically consist of a septic tank and a drain field. A system failure can occur when soil absorbent capacity is exceeded by the effluent. Other causes for system failures include the use of poor design of septic systems, poor construction, mismanagement and poor maintenance, and poor soil conditions. Note that risk is primarily from direct runoff of partially treated effluent where the drain field or treatment system has failed or an indirect release in areas where impacted groundwater recharges surface water. Some of the potential pollutants sources from failing tanks include nitrate contamination, nutrients, toxic chemicals, organics, pathogens, and PCPs.

Wastewater Treatment Plants

They are classified as moderate-higher risk factors on the quality of water. PCPs that go through the wastewater treatment process are emerging contaminants of concern. Wastewater bio-solid and recycled water application areas may impact surface water through direct runoff or if impacted groundwater recharges surface water. Another separate, but related risk are sewer lines that are in close proximity to water sources or are located in highly permeable soils adjacent to water sources. These sewage lines should be considered a high priority for evaluation of potential leakage, and as a preventive measure should be a high priority for regular inspection and maintenance.

Summary of Risks-according to World Health Organisation (WHO) view point, the highest risks to water quality in most sources is storm water runoff characterized below as;

- Urban storm sewer discharges
- Storm water runoff impacts due to growth and development in urban areas (for instance, conversion of farm and forest land to urbanized development),
- Storm water runoff and potential spills emanating from various industrial and commercial facilities
- Storm water runoff from agricultural practices,
- Storm water runoff from Roadside vegetation management,
- Potential domestic wastewater discharges from wastewater treatment plants and septic systems, and
- Hazardous material spills from commercial and industrial areas, transportation activities along major highways , the railroad line, and the numerous road bridges that cross big rivers in urban centres

WATER QUALITY AND WATER QUANTITY ASSESSMENT

There are five key indicators that any assessment action plans for water quality and water quantity should focus on. These indicators include;

- Nutrients in water
- Bacteria in water
- Pesticides in water
- Water temperature
- Sediments in the water

When assessing these indicators, they should be ranked based on their potential impact on the quality of water for human consumption. That is low risks, moderate risk, or higher risks. Additionally, each indicator should be ranked based on its concentration in the water. Although not all of these five water quality indicators are considered to be potential risks, their assessment can be used to prioritize the most appropriate mitigation strategy that should be taken.

Different countries across the world have set programs, regulations and rules to govern and protect the quality of their water sources. Most of the utility companies and other governmental agencies in specific countries usually have primary roles in protecting the quality of water for the citizens. These roles primarily falls under two categories, which are management and control of water resource usage and the second category entails environmental management and protection. These measures are primary focussed on restoring and protecting the quality of water for human consumption, with other secondary benefits on protecting aquatic life. The governments of specific countries can support these proactive measures in protecting water resources through the following;

- Supporting the regulatory efforts: this can be done through promoting protection actions taken on water quality and specifically addressing drinking water quality
- Tracking changes in regulations: this can be done through monitoring various regulation and rules enacted to deal with impacts on drinking water quality
- Advocating for more protections for the source of our drinking water or prioritizing existing resources toward the drinking water resource and public health goals.
- Setting pollution goals or targets: this can be achieved through working with other governmental and non-governmental regulators and partners, including public and private water services companies, the department of agriculture, department of forestry services, state's department of transportation, department of environmental quality, department of health and human services.
- Protecting large surface water sources that are also drinking water sources. This can be done through monitoring various regulations and rules enacted to ensure no changes are made that could negatively impact the water quality in different water resources
- Tracking and influencing land use changes at community and national level such as comprehensive plan development or changes in riparian ordinances, septic code, and open space or floodplain development/restrictions.

The general strategy used in protecting the quality of drinking water should include elements that address different measures, strategies and programs that need to be taken in water management in order to address different risks and threats to the quality of water and to ensure that there is long-term viability in water resources used in a country. Some of the key strategies that can be used in protecting the quality of water are discussed below.

WATER RESOURCES ANALYSIS

This strategy can be undertaken in order to identify and understand better on various potential risks and threat factors on water resources. Under resource analysis strategy, there are four key elements that can be focussed on; these include additional studies conducted by governmental and non-governmental bodies, the use of Geographical Information System (GIS) in order to identify an spatially localise potential risks and threats on land use, pollutant load modelling, and finally, the use of monitoring programs in order to comprehensively assess the quality of water in different water sources.

Governmental and Non-Governmental Additional Studies

These studies mainly provide the water providers with valuable information regarding the watershed in general, the quality of the water source and flow. Additional studies would help inform decision making. These studies, listed below, are not prioritized. They include the following:

Taste and odour/algae studies-although not normally considered as direct drinking water threats, taste and odour events are a nuisance and usually affect water quality in one way of another. Algae can clog intake structures requiring a higher level of maintenance, and can occasionally prevent water systems from withdrawing water. Studies have shown that algae in most parts of the world may clog a water intake structure, reducing its capacity to withdraw water. The purpose of such a study should be to investigate algal growth and bloom in major water sources and associated issues of taste and odour and intake clogging. Taste and odour/algae studies can be used to determine if algal blooms are associated with PGE operations or if they are associated with other human activities such as agriculture, forestry, urbanization, and industrial development. By evaluating nutrient sources it would help in development of prediction and mitigation strategies.

Groundwater Study – The purpose of this study would be to develop a better understanding of groundwater/surface water interactions within the watershed. If areas of high groundwater/surface water interaction were identified this would help prioritize upland point and nonpoint pollution mitigation strategies. This would include looking at old abandoned wells (domestic, agriculture, and geo-tech) as potential sources of pollution.

Additional Flow Balance –a study on groundwater wells could provide indication in the timing and spatial distribution of groundwater discharge to the lower sides of major water resources (such as rivers and streams). The purpose of additional flow balance study is assess the stream flow gain/losses in the lower catchment areas of a river, to track surface-water sources to (or sinks from) the lower catchment areas, and to estimate the quantity of groundwater discharges in to the system.

Climate Impact study –The implications of climate change are considered as a crucial area that needs more study to better understand how climate change may affects water supply over the decades to come. Based on climate change models it can be predicted that there will be warmer winter temperatures and winter precipitation in the form of rain instead of snow in the Pacific Northwest regions. This is likely to lead to declining snow packs and reduced summer stream flows. The impact could be increased turbidity and flood events in the winter and degraded water quality due to less dilution during the summer months. Although regional modelling has been done in most countries, there has been little climate work done in various developing countries, and this acts a precipitating factor towards poor quality water in future. This kind of study would help a country develop adaptation and mitigation strategies to deal with potential changes in water quantity and water quality due to climate change in various water sources. North East South America may face drier climates by the second half of the 21st Century according to some recent modelling weather forecast studies.

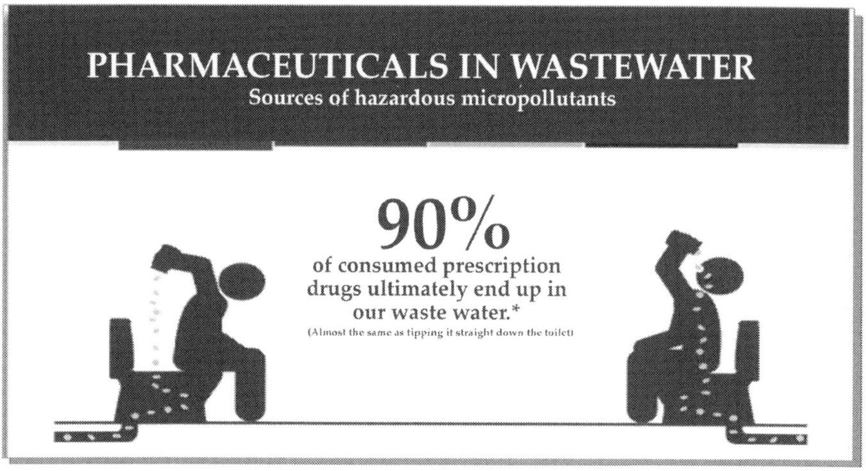

Figure 6.3: The risk of pharmaceuticals and personal care products on water quality.

Source:*http://micropollutants.com/About-micropollutants*

Pharmaceuticals and Personal Care Products (PPCPs) – there are different pharmaceutical products and by-products that have been identified in various surface waters, treated sewage effluents, soil, and tap water at extremely low levels (parts per trillion). Science currently shows that these levels are unable to induce acute effects in humans, as they are far below the recommended prescription dose, but they have been found to affect aquatic ecosystems in one way or another. Although there currently are no drinking water standards/requirements for PPCPs there is growing public concern around this issue. It may be in the country's best interest to learn more about these emerging contaminants and to develop a water quality sampling program in anticipation of future regulation

Geographical Information System Assessment (GIS)

Conducting a GIS analysis of water risks, as mentioned above, is used to provide spatial context of where the risks are located as well as their intensity (numbers of threats clustered together rather than spread out over a larger area). The GIS risk layers produced can then be overlaid with sensitive soil layers and/or proximity to tributaries which will identify potential hot spots.

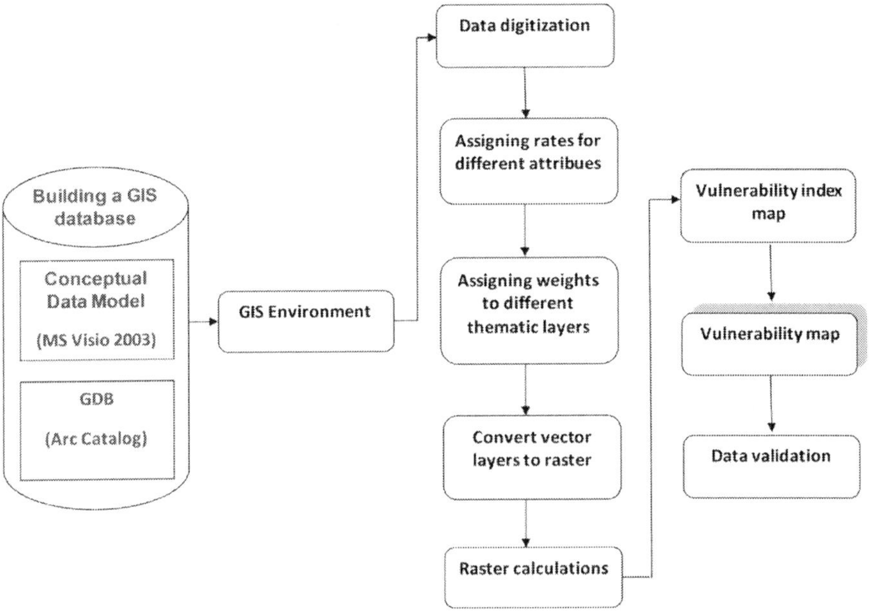

Figure 6.4: Application of Geographical Information System assessment in water quality.

Source:_http://file.scirp.org/Html/6-8401229_33275.htm_

The purpose of Geographical Information System assessment is to help in prioritizing mitigation programs. The production of a detailed drinking water risk map is a powerful tool for the government to prioritize source protection actions, seek grant funds, and bring in additional partners for projects on the ground. It also acts as a visual tool to communicate with other agencies and organizations that are interested in partnership and also allow the government to begin a dialogue with industry and stakeholders to collaborate on future projects. It could potentially help gain more interest from federal, state and local agencies, watershed councils and others to conduct studies that will help in collecting, processing and analysing data within the watershed and have a better understanding of what needs to take place in order to protect water sources. Some of the proposed studies that can be conducted under Geographical Information System include;

Scoping Study: A scoping study includes an outline/concept of the data needs and data gaps, planning/assessment capabilities, and software/modelling options to develop a flexible and adaptable monitoring/

modelling program. To effectively map all of the drinking water threats in Geographical Information System there are several steps that must be undertaken, and they include;

- Identification of what Geographical Information System data would be needed, what existing data is available, potential partners to acquire the Geographical Information System layers from, as well as identifying software/hardware needs.
- Laying out how Geographical Information System development that would integrate with water quality monitoring and pollution loading modelling work identified under the water protection plan.
- Determining where the Geographical Information System data will be housed and maintained

This scoping study will also help refine the budget estimate on what it will cost to perform the Geographical Information System work. In addition the study may also help in identifying partnerships that could help fund the project, whether with dollars or through in-kind services. This study may also include scoping and data needs for pollution load modelling to identify which kind of pollution load model is the most appropriate for the watershed and to refine the costs of doing such modelling.

Geographical Information System analysis support for assessing threats to water quality from septic systems: This can be done through cluster analysis in order to identify those areas of highest potential risk to water quality from septic systems. Additionally, septic system age and soils data can be combined with the results of cluster analysis to identify relatively high density septic system clusters, in proximity to surface water, in areas of rapidly draining soils. This analysis is important especially in assessing the threats to water quality from septic systems as well as help in targeting further monitoring efforts.

Geographical Information System analysis supports for assessing threats to water quality from agriculture activities: this entails obtaining portions of the river watershed that contain significant areas of agricultural production. These portions may be acquired by Geographical Information System staff from different agricultural unit (field) locations and crop type. Assumptions on pesticide and fertilizer application rates should be appended to Geographical Information System data set in order to assess potential effects of agricultural use on water quality and to facilitate further analysis and/or water quality modelling at a future date.

Geographical Information System analysis support for assessing threats to water quality from forestry activities: Forest cutting and herbicide/fertilizer spray application notices pertaining to properties within watersheds should be acquired from department of forestry. Cutting and spray areas should be digitized into Geographical Information System with tabular information from the notices entered into a forest spray tracking database. Over time this can be used as a means of assessing concentrations of pesticide applications on forestland and their potential threats to water quality.

Geographical Information System analysis support for assessing threats to water quality from urban development: country plan designation and tax assessor Geographical Information System parcel data should be combined with selected development constraints layers in order to develop a potential residential "build-out" capacity estimate, in terms of residential units, for water source.

Geographical Information System analysis support for assessing threats to water quality from vulnerable soils: An analysis of vulnerable soils should be conducted and includes layers to evaluate erosion, runoff, landslides as well as sensitive areas such as wetlands, floodplains and water setbacks. This may also include the collection of the high spatial resolution Light Detection and Ranging (LIDAR) data from Department of Geology and Mineral Industries in order to identify areas where historic landslides could be reactivated by changes in land use practices, road building/alteration, and development.

Watershed/Pollution Load Modelling

Water quality and biological data collected through the water quality monitoring program should be used together with GIS threat data layers to build a watershed model that allows prediction of rainfall-runoff relationships, pollution load calculations, contaminant transport scenarios, and impacts due to changes in land use. This tool helps in analysing watershed data in a way that provides graphic results that demonstrate watershed health and allows the government to focus on areas of potential water quality degradation. Land use changes should be run in order to assist in prioritization of efforts on water source protection and to better understand long-term threats and trends in water catchment. Pollution load modelling can also be used to determine what kind of water quality data is needed to support future water quality monitoring efforts. To effectively do this, the government of a country should look at the different kind of pollutant loading models that are available. Pollutant load model for a small portion of the lower basin can possibly be expanded to cover the whole watershed. This may lead to some partnering of resources. As part of this process it is essential to assess the data available to drive the hydrology modelling in the basin, as well as determine if additional rain gages are needed.

Water Quality Monitoring

This is one of the key elements of the water resources analysis program. There are four key issues that can be addressed using water quality monitoring. These include the following.

- Provision of a baseline that can observe water quality trends over time
- Verification of the identified hot spots using GIS mapping
- Evaluating the relative success of restoration, other protection strategies, and Best Management Practises (BMP)'s over time
- Identification of existing water quality monitoring gaps

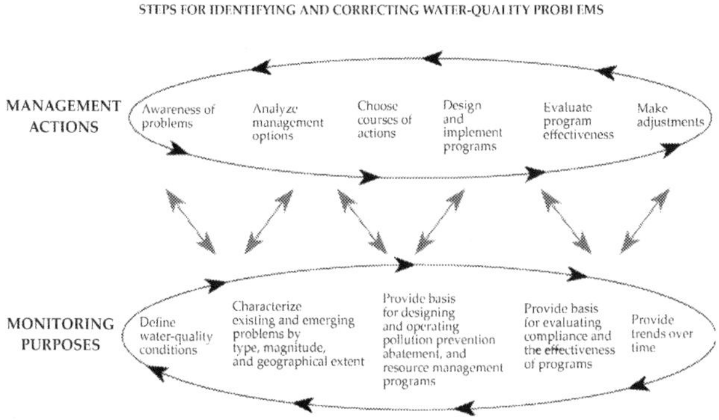

Figure 6.5: Water quality monitoring and management model.

Source:*https://acwi.gov/lopez.main.html*

Addressing these four issues allow water providers to evaluate water quality, biological health, and land use trends to pinpoint adverse trends that can result into drinking water threats. This information may help in prioritization of problems or threats to the drinking water source early on and target source water protection efforts based on the data. Water quality monitoring data can be used with GIS and together with hydrologic pollution and simulation loading models. To successfully achieve this, the following must be done;

- Encouraging and supporting existing monitoring programs for collection of baseline data on a fixed interval monitoring (i.e., monthly, quarterly)
- Encouraging more storm event monitoring
- Identifying all of the existing monitoring programs in the watershed and various gaps where additional water quality sampling and monitoring are needed
- Once monitoring gaps have been identified, additional monitoring approaches should be developed in order to address these data gaps. These could include the following:
  a) Developing a water quality monitoring plan for the whole basin

  b) Developing a cooperative water quality monitoring plan and program which could include developing a storm event monitoring program for the area outside the service area. This would include identifying funding sources for water quality monitoring and/or writing agreements with potential partners to do the work
- Developing a storm event monitoring program for a selected suite of analyses for laboratory analysis based on the land use and threat being evaluated
- Implementing a periodic water quality monitoring program that targets specific pollution sources such as pesticides, pharmaceuticals and personal care products, disinfection by-product precursors, or hydrocarbons
- Implementing water quality monitoring that verifies "hot spots" contamination potential

- Implementing a Macro-invertebrate sampling program to establish a baseline for lower river sub-basins water quality and resample periodically
- Developing an information management tool to support the collection and management of water quality data

RESEARCH AND EDUCATIONAL PROGRAMS

Elementary and High Schools

The main aim of research and educational programs is to promote and enhance awareness of different people on issues related to water quality and water sources protection. These programs can be done in communities, schools, health institutions, and events in religious facilities. Some of the key water issues include cultural issues affecting water supply, biological issues, and other socio-economic issues that might affect the quality of water in different parts of the world. The governments of specific countries should take the initiative of incorporating specific water programs in school curriculums in order to promote awareness of young generation (students) on watershed related issues, water quality issues, environmental quality of water, and the health of different water sources (such as the rivers, lakes, streams, and reservoirs) among other issues.

Youth educational programs can be introduced in schools water sciences, biology, and chemistry programs in order to encourage teachers to conduct water quality, macro-invertebrate, and bio-assessments survey work on various water resources as a part of their on-going curriculum. This will provide students with a comprehensive educational experience regarding water quality, aquatic habitat, stream morphology, riparian functions, and laboratory analysis associated with the health of watersheds. The government should work with teachers on procedures to make sure the collected data can be used to model pollution loadings, conduct trend analysis, and identify problem areas. This further benefits the students as they become involved in how their work and data are applied to solve real life problems. The government should also pursue working with various researchers conducting water related projects (including teachers, middle and high school students, scientists, businesses, governmental agencies, community groups, and metropolitan schools in the country).

Graduate Schools

It is also important for respective governments and other non-governmental agencies to work with local universities in order to promote research and graduate work in the watershed. This can be done through identifying issues or problems in the watershed that graduate research could provide direct benefit, communicate with universities to find graduate or undergraduate programs that match the issues or problems, and offer research grant funding for universities to conduct research.

Other areas of potential study: There are a number of areas that will require additional study that could utilize local universities or other governmental and non-governmental partners. They include additional flow modelling on the lower basin, ensemble modelling, climate change studies, carbon studies, taste and odour/algal studies, ground water studies, additional flow balance studies, studies looking at the difference between treatment processes, how hydrocarbons get into the water, macro-invertebrate studies, and land use change modelling.

EVALUATION AND MITIGATION PROGRAMS ON WATER POLLUTION POINT SOURCES

The main aim of these programs is to help in tracking, assessing, and monitoring various water point sources in order to mitigate various potential pollutants that affect water quality. To execute and achieve the best results from these programs, it is important that other stakeholders (such as regulatory agencies, permits, and other governmental utility facilities) to be involved. When executing these programs, it is advisable to look at different permits that are issued by different facilities such as water pollution control facilities, wastewater treatment facilities, sand and gravel mining, fish hatcheries permits, storm water permits, confined animal feeding operation permits, PGE dam licences and permits, and air contaminant discharge permits. Some of the major risks associated with these kinds of point source or permits include spills, leaks, and various contaminants released to both surface and groundwater.

Figure 6.6: Major point sources of water pollution.

Source:*http://www.lakescientist.com/pollution/*

Tracking and evaluating these permits will help in identifying potential water pollutants. Once the water threats are identified, mitigation actions should be taken based on the type of facilities responsible. Mitigation targets should be set based on practices for water management, requirements of specific permits, reports on permit compliance, monitoring results obtained from various water quality programs, and information on permit holders. There are different strategies that can be deployed in order to mitigate impacts or threats to the quality of water. These strategies include;

- Reviewing and commenting on permits to see if there could be an impact on the quality of water
- Identifying ways to enhance permit requirements.
- Improve monitoring of waste discharge from different facilities
- Providing incentives for businesses that recycle, reuse, and reduce various waste streams
- Developing a recognition program for those business that take voluntary actions to improve water quality

Wastewater Treatment Plants Permits

There are several domestic wastewater treatment plants that discharge effluent into different water sources in most parts of the countries across the world. Some of the key mitigation activities that can be taken in order to control and prevent threats on the quality of water include the following;

- Reviewing water quality permits for wastewater treatment plants as well as the supplemental management plans for bio-solids and recycled water
- Improving notification protocols for wastewater spills and sanitary sewer overflows by working with various responsible facilities to review the protocols
- Providing watershed residents outreach materials with information on proper pharmaceutical and household hazardous waste disposal methods as well as reduction of overall toxic chemical use
- Supporting various events undertaken by various governmental and non-governmental agencies to dispose of pharmaceuticals

Fish Hatcheries Permits

Different countries around the world have specific permits given to various fish hatcheries with regards to waste discharge into water source such as river, streams, and lakes. To effectively control and prevent different threat to the quality of water, these are some strategies that can be deployed with regard to these types of waste discharge permits. These include;

- Determining if there is the need for hatchery effluent monitoring for nutrients, BOD and pharmaceuticals (antibiotics). Based on monitoring results, management options to recycle water and otherwise minimize the release of hatchery water may be considered.
- Contacting permits operators to talk to them regarding the location of their systems within a country's drinking water protection area as determined by relevant water utility companies, and provide them with educational and technical

resources to encourage them to use best management practices

- Contacting the responsible facilities in order to inquire about the status of existing permits for identified facilities and if they are in compliance or not
- Reviewing various fish hatchery National Pollutant Discharge Elimination System permits including reviewing the facilities' effluent discharges and chemical use to see if these could impact drinking water quality.

Sand and Gravel Mines Permits

Studies have shown that mines contribute hugely towards water quality contamination. There are different permits issued to various mine facilities in order to control and prevent water quality contaminations. Some of the crucial strategies that can be deployed in order to control water pollution by sand and gravel mines include;

- Identifying mines and gravel pits within different watersheds in order to determine if they are active or not
- Reviewing National Pollutant Discharge Elimination System permits on mining. This should include reviewing the facilities' discharges to see if these could impact drinking water quality.
- Contacting the Department of Geology and Mineral Industries in order to ensure that permitting staffs are aware of various water sources before authorizing new mining permits. Request they do compliance inspections or provide technical assistance to mines in areas that are considered critical for drinking water protection.
- Contacting mine owners and operators to discuss the location of their mines as it relates to the drinking water protection area. Educational and technical resources could be provided to encourage them to use best management practices.

Storm Water Permits

There are different storm water permits applicable in different countries across the world. There are general storm water permits held by businesses and industry. Unpermitted storm water should be addressed in regard to non-point source section of the governmental plan and many of the mitigation strategies for permitted or non-permitted storm water should be addressed. These include;

- Working with relevant authorities in order to understand how they are implementing their storm water permits and where the government can provide encouragement or pressure to do better.
- Encouraging various industries and businesses to reduce and/or cap storm water pollution
- Encouraging all cities in the watershed, to periodically conduct dry weather field screening at all storm sewer outfalls in order to detect and control illicit discharges
- Encouraging responsible authorities to conduct more frequent inspections of commercial, multi-family residential and industrial facilities to control illicit discharges and to improve storm water runoff quality
- Identifying individual facilities or areas with a concentration of facilities with storm water permits, and encourage them to construct on-site and/or off-site treatment systems or buffers (e.g., wetlands, bios-wales, retention ponds, etc.) and identify resources and/or partners that could help provide incentives to get these kinds of treatments implemented
- Contacting permit holders to encourage them to reduce their storm water pollutant concentration

Water Pollution Control facilities Permits (WPCF) for Domestic Wastewater

These include all the larger permitted septic sewer systems. Some of the effective strategies that can be deployed in order to enhance compliance of these permits in controlling and protecting water quality include;

- Reviewing and commenting on WPCF permits for any impact to drinking water quality
- Contacting permit operators to discuss the location of their systems as it relates to the country's drinking water protection area. Educational and technical resources can also be provided to encourage them to use best management practices

- Mitigation strategies listed for smaller non-permitted septic systems should also be appropriated for these larger systems

Confined Animal Feed Operation (CAFO) Permits

There are different permitted CAFOs within different watersheds in specific countries across the world. To effectively control and protect the quality of water in these watersheds, there are a number of strategies that can be taken. These include; Reviewing and commenting on CAFO permit renewals, as part of the public comment process, to make sure that Animal Waste Management Plans are protective enough on safe water supply and contacting individual CAFO owners to ensure that their emergency response programs establish adequate spill notification methods with regards to drinking water supply.

EVALUATION AND MITIGATION PROGRAMS ON WATER POLLUTION NONPOINT SOURCES

The main aim of these programs is to enhance tracking, inventory, evaluation, and monitoring initiatives on water pollution nonpoint sources. Once these sources are identified and evaluated successfully, mitigation actions should be taken in order to control pollution for safe and clean water supply for all. All the responsible authorities should define and provide a better understanding about nonpoint source threat and offer potential strategies on how to curb and control these nonpoint water pollution threats. Some of the major nonpoint source water pollution contributors include rural and urban storm water runoffs, agricultural activities, and forestry activities. It is also significant to understand the effects of septic systems and other recreational uses (including floaters and boats) and other nonpoint source water pollution in watersheds as potential threats to water quality.

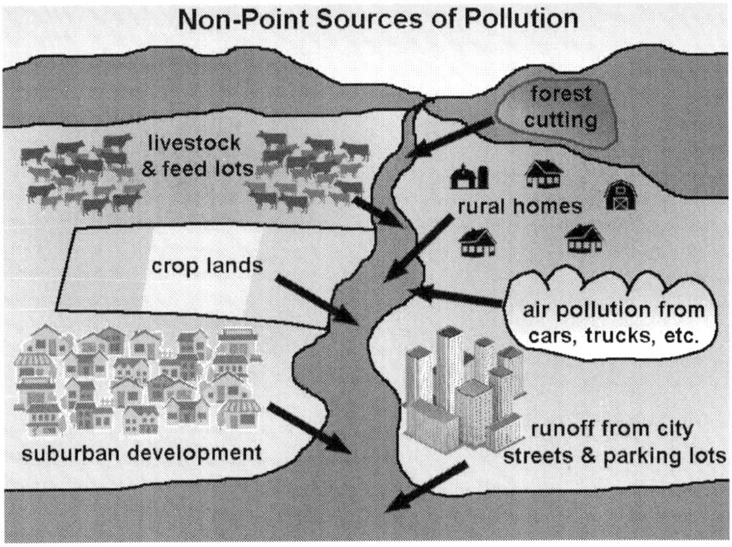

Figure 6.7: Illustrating Key Non-point sources of water pollution.

Source:*http://gotbooks.miracosta.edu/oceans/chapter17.html*

To effectively control and manage these nonpoint source water pollution threats, it is advisable that all the responsible authorities, agencies, business owners, and land owners to work together in implementing and practising various management strategies to mitigate potential water quality threats in different parts of the country. There are several tasks with regards to controlling water quality threats from nonpoint sources. Some of the key categories of mitigation strategies include the following;

Inventory reviewing: This entails reviewing and updating various assessments of potential nonpoint source areas of water pollution within different watersheds. Information regarding locations of nonpoint source pollution can be stored and obtained if needed in the geographical information system sites. This information can be used for tracking and analysis activities.

Tracking and evaluation: this entails evaluating and tracking various activities associated with nonpoint source pollution such as pesticide use and application schedules, crop types and cycles, forest management activities, building permits with septic systems, evaluation of high risk areas due to septic system density, age and past failure rates.

The comprehensive monitoring program can be used to provide good information on the actual impacts from these activities. This information can be tracked and evaluated using geographic information system in order to allow respective governmental and non-governmental agencies to identify high-risk areas (areas that are already impacted as well as those at risk for impact).

Mitigation actions: With regard to the information obtained concerning potential threats, the type of nonpoint water pollution source, and other information, hot spot areas should be targeted for mitigation. Hot spots are defined as sensitive areas such as flood plains, highly permeable soils, thin soils, high slopes, areas in close proximity to various watersheds or in areas where risk is clustered (numerous risks in a small geographical area rather than spread out). There are different options and strategies that can be taken in order to mitigate these nonpoint water pollution threats. These include; increasing monitoring activities regarding water quality in those highly affected and potential areas; encouraging various authorities, agencies, business owners, and land owners to implement best management practices; and identifying resources (technical assistance) and/or partners that could help provide incentives (funding or grants) to get these kinds of treatments implemented. Some of the nonpoint sources of water pollution include the following.

Unpermitted Urban Storm water Runoffs

These are key threats to various water sources in a country. Most unpermitted urban runoffs usually contain harmful substances that can threaten the quality of water for human consumption. These substances include pesticides, bacteria, heavy metals, petroleum hydrocarbons, fecal materials, sediments, different nutrients, chlorides, and other different kinds of organic compounds. There are several mitigation practices that can be deployed in order to control and reduce the impacts of these substances on various water resources. These practices include practising proper designs for drainage systems, using low impact development practices, best management practices, and using state of-the-art storm water management techniques. The governments should work closely with other organizations and private agencies to properly analyse and develop specific guidelines and strategies to mitigate storm water runoffs in various urban centres of the country. Some of these strategies include;

- Encouraging partners to enhance implementation of erosion and sediment control programs. This can be done through technical assistance, providing funding or identifying other resources
- Educating the watershed community about the impacts that storm water runoff has on the water quality
- Monitoring storm water and urban runoff and provide support on implementing plans with regards to storm water runoff management
- Striving to obtain reduced storm water volumes and pollutant levels in the watershed's urban areas. This should include developing incentive programs to encourage voluntary implementation of these types of best management practices.
- Encouraging wider and more fully functioning riparian buffers and the implementation of riparian best management practices to enhance the function of degraded buffers
- Promoting and providing technical assistance regarding sustainable and low impact development practices (such as green streets) within various cities and big town areas to reduce storm water discharge volumes and pollutant loads
- Facilitating workshops geared for parks, maintenance facilities, nurseries, etc. on chemical applications, pollution prevention and waste reduction, and proper chemical use, storage, and disposal.
- Supporting (financially or otherwise) various river basin council efforts to enhance riparian areas in the vicinity of these facilities or permits to increase filtering capabilities and stream health
- Identifying commercial and industrial customers with large manicured landscapes including golf courses and state, county, and city parks and facilities within watersheds and work with them to decrease fertilizer and pesticide use and to implement Integrated Pest Management methods. This should include facilitating and/or providing training for these users

Rural runoff in Unincorporated Areas

Run off in unincorporated residential rural areas can lead to turbidity, erosion, and bacteria that can impact water quality. To better understand these potential impacts the responsible governmental and non-governmental agencies should:

Develop education and outreach materials for rural residents including: promoting proper household hazardous waste and pharmaceutical disposal; encourage best management practices for pesticide use, storage and application, aboveground and underground storage tank operation; maintenance and leak prevention; private well maintenance; septic system operation/maintenance; and animal waste management. They should also find out if private wells are a conduit to groundwater pollution in areas where groundwater recharges surface water. Encourage well registration and proper well maintenance. This could also include incentives to ensure proper well abandonment.

Agricultural Mitigation Programs

Agricultural is considered as key nonpoint source of water pollution in most parts of the world. There are different actions and strategies that can be deployed in order to mitigate threats on water quality emanating from agricultural activities. These include;

- Providing technical and/or financial assistance to help various responsible agencies to carry out implementation of cater conservation programs. Such programs include; Agricultural Chemical Removal Program (the purpose of this would be to help growers remove and properly dispose of old agricultural chemicals that still may exist in barns and sheds), Organic Farm Certifications, Pesticide Reduction Program (this includes targeting pesticide education for urban and rural lower basin landowners, encouraging nursery and agricultural industry to reduce pesticide application, and encouraging alternatives to pesticides), Forestry Mitigation Programs (forest management activities such as the use of chemical applications for stand treatment, road building, and various timber harvest techniques have the potential to adversely impact downstream water quality)

- Identifying and mapping various crop types in water catchment areas using aerial photography, geographical information systems, and ground-truthing in order to come up with an effective baseline inventory of various agricultural activities in watershed areas
- Coordinating with the country and cities on urban forest planning
- Analysing various types of crops, the type of pesticides and fertilizers used on crops, the type of soils, and the proximity to various surface and ground water bodies that may be affected by different storm water runoffs contaminants from agricultural activities
- Conducting a windshield survey of agricultural lands and informing relevant agencies about potential water quality concerns within drinking water supply areas
- Determining if additional water quality monitoring would be appropriate in high priority areas. Also determining the ideal frequency for additional pesticide monitoring
- Identifying areas where buffer strips (or riparian restoration) could be installed or enlarged to capture runoff from agricultural lands
- Mapping all timber harvesting activities in various national forests as well as on private timber lands that occur in the watershed in the past five years and those that are anticipated to happen in the next five years. In addition to mapping timber harvests, identity if chemical spray treatments were/will be applied. Land owners, however, should be required to submit a paper copy with the application so the information is available but would have to be collected and digitized for a GIS layer
- Providing agriculture landowners with information on recycling and reusing water from irrigation as well as look at agricultural ponds for potential impacts on water quality
- Working with agricultural partners to prioritize technical assistance efforts in areas that are higher risk to the drinking water supply
- Conducting an analysis of the creek basins that received aerial

applications of pesticides or herbicides to highlight the basins that had higher amounts of chemicals applied

- Determining if there is a need to begin a storm event monitoring in creek basins that have a higher risk of chemical runoff due to recent stand treatment and harvest activities

Septic System Assistance Program

Most septic systems in big cities and urban areas that are poorly managed and maintained can act as potential nonpoint source for water pollution. Studies have revealed that these septic systems usually release high levels of synthetic organic substance, bacteria, organic matters, toxic metals, viruses, ammonia nitrogen, personal care products, and other pharmaceutical products to various surface and groundwater bodies. Long-term solutions should be developed with respect to various developers, contractors installing septic systems, residents around watershed areas, and other regulatory governmental and non-governmental agencies. All these responsible stakeholders should try to minimize threats emanating from septic systems through; Overlaying information on septic system age and sensitive areas such as Federal Emergency Management Agency (FEMA) flood plain maps, soil/geology maps for highly permeable, thin soils, high slope, proximity to ditches, streams or rivers which could be used to determine high risk areas; targeting high risk septic system areas for further outreach efforts, technical assistance, or incentive programs; determining if additional water quality (surface/groundwater) sampling is necessary and if so try to develop a plan and implement approach to conduct water quality monitoring of septic systems impacts; providing homeowners with technical or financial support through governmental and non-governmental funding and/or grants to inspect and pump-out (if needed), repair and/or upgrade septic systems that have been identified in high risk areas; offering workshops with easy to understand information about maintenance of systems and options/recommendations about the need upgrades and/or replacement of old on-site system that are failing or near failure; encouraging and advocating for state-wide or county enhanced regulation of septic systems (i.e. maintenance schedules, required inspections on property sale; and encouraging administering agencies to adopt a more stringent proactive septic inspection program rather than the current, complaint driven reactive approach.

Recreation Impacts

Most watersheds in a country have a variety of recreation facilities including forest service and PGE campgrounds, county and state parks and boat ramps, as well as a number of golf courses. Recreational activities also include hiking, rafting, kayaking, fishing, hunting, and floating. Potential contamination from these activities include: petroleum products, solvents, paints, oils and grease, and untreated or partially treated human waste (nutrients and pathogenic microorganisms). Sedimentation can also be an issue resulting from stream bank erosion from boaters, anglers, off-trail hiking, or Off Highway Vehicle (OHV) activities. Golf courses and parks may use herbicides, pesticides, fungicides, and fertilizers. To address these potential impacts should be done:

- Advocating for the creation of boat no-idling zones around drinking water intakes
- Monitoring for fecal coliform at two or three strategic locations relative to high use/impact areas
- Supporting programs that provide and maintain restrooms and portable outhouses for river users
- Encouraging boaters to implement environmental practices while boating, to help protect water quality
- Providing free spill kits (such as absorbent pads) for boats
- Educating users and golf course landscape managers about the most effective means of reducing impacts on water quality from various recreational activities

CHAPTER 7

BENEFITS OF REUSING RECLAIMED WASTEWATER

INTRODUCTION

Statistically more than 15,000 Publicly Owned Treatment Works (POTWs) in U.S. were in operation by the year 2000. This made it possible to treat more than 40 billion gallons of reclaimed wastewaters on a daily basis. More than 95% of POTWs were owned by different municipals by the year 1996, which helped in collecting, treating and disposing wastewaters from more than 180 million citizens of United States. Currently, more than 1 billion gallons (or 2.5%) of commonly referred to as reclaimed wastewater that is treated on a daily basis is used as non-potable water for different purposes such as irrigating various golf courses and other publicly owned parks in United States. Compared to the reclaimed wastewater treated and used in other countries such as Israel, wastewater that is treated in U.S. is significantly low. For instance, Israel accounts for more than 75% of its wastewater that is treated and reused in various purposes.

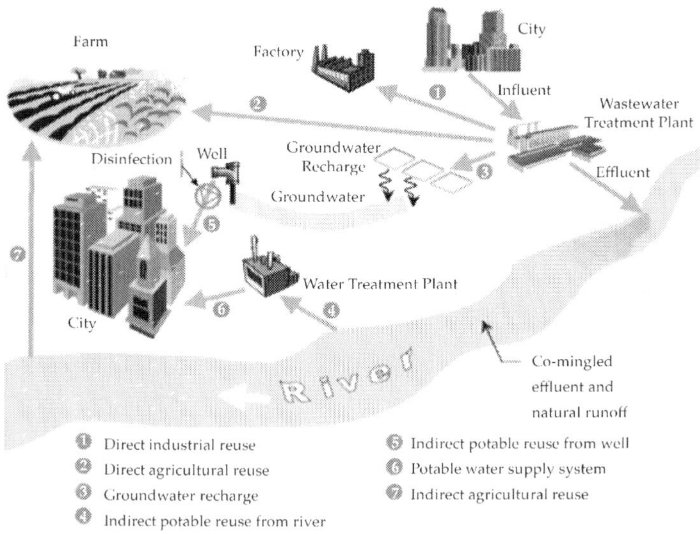

Figure 7.1: Direct and Indirect Recycled Water Use.

Source: http://web.mit.edu/12.000/www/m2012/finalwebsite/solution/waterreuse.shtml

In 2000, countries such as Florida, Texas, and California accounted for more than one third of the total freshwater withdrawals. The highest withdrawals of surface freshwater sources were from California. Some of the key sectors that consumed the highest volume of these freshwater include thermoelectric power production sector and agricultural sector. The highest volumes of groundwater withdrawals were also made by Nebraska, Texas, and California, with the sector that accounted for the highest withdrawal been agriculture. If these countries used more of their reclaimed wastewaters for agricultural activities, more of the surface and groundwater sources could be conserved thus preventing the current enormous depletion rates. For instance, it was projected that the demand for freshwater for more than 45% of the 15 million additional citizens in California would be met by 2030. Some of the key areas that reclaimed water can be used include toilet flushing, for irrigating various landscapes, and for recharging various aquifers in most parts of north western parts of America. Water depletion problems can also be solved through the use of reclaimed wastewaters (either directly through recharging the groundwater basins or indirectly through reusing).

PLANNING FOR RECLAIMED WASTEWATER REUSE

Effective planning is a key aspect in the reuse of reclaimed wastewater in most parts of the word. All the involved sectors and stakeholders should come together and focus on consolidating various agendas into a single goal. Reuse of wastewater is usually considered as both iterative and interactive process that requires proper planning to achieve a specific objective and goal for implementing the project of water reuse. This plan should consider all the project areas, potential markets, and capitals to execute the project. There are certain elements that should always be considered when planning for reclaimed wastewater reuse project. These elements include;

- Assessing all the needs for reclaimed wastewater treatment and disposal
- Assessing all the demands and distribution systems
- Assessing all potential benefits that may be acquired through reuse of reclaimed wastewater
- Analysing all potential markets for reclaimed wastewater
- Engineering and economic analysis
- Analysing all the financial plans for implementation process
- Assessing programs for public information

The process of planning for reclaimed wastewater reuse involves three key stages. These include (1) the conceptual planning stage, (2) the preliminary feasibility stage, and (3) facilities planning stage.

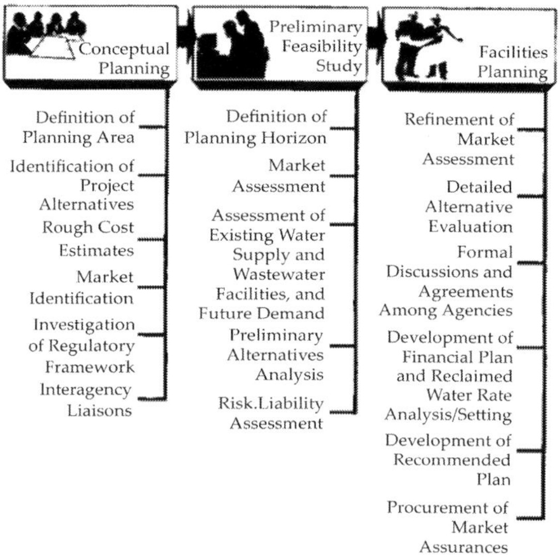

Figure 7.2: Project Planning Phases.

Source:*https://www.dsr.wa.gov.au/support-and-advice/facility-management/developing-facilities/feasibility-study*

CATEGORIES OF WASTEWATER REUSE

For any reclaimed wastewater to be utilizable, it has to undergo a process of treatments. Usability of wastewater highly depends on the quality of the effluent. There are four key categories wastewater reuse which include the following:

Direct and Indirect Potable Wastewater Reuse

This is the first category of wastewater reuse that represents the highest quality. Potable wastewater is directly blended from treatment facilities into the water supply system. Indirect reuse of potable reclaimed wastewater involves augmentation into water supply system. Sometimes reclaimed wastewater is used indirectly when it is discharged into upstream of the river. This is commonly referred to as "unplanned indirect potable wastewater reuse". Because of the high public safety requirement for water quality, most countries do not practise direct and indirect potable wastewater usage.

Recreational and Unrestricted-Access Urban Wastewater Reuse

This wastewater should be of high quality too. some of the reuses of unrestricted-access and recreational urban wastewater include irrigation of sport and parks fields, fire fighting, in-building uses such as air conditioning and toilet flushing, and decorative fountains.

Recreational and Restricted-

Access urban wastewater: people are mainly restricted to come into direct contact with reclaimed wastewater in this category. Some of the most common reuses of wastewater under this category include controlling dust, irrigating landscapes, and other municipal works uses.

Industrial Wastewater Reuse

This wastewater reuse category plays a very vital role in meeting both public and economic safety concern. Typical uses of industrial wastewater include concrete manufacturing, aggregate washing, washing equipment, tower cooling, scrubbing stacks, feed boiler, water processing, and other construction activities.

PUBLIC CONCERN AND RISK ASSESSMENT ON RECLAIMED WASTEWATER

Those areas where reclaimed wastewater is rarely reuse, there is usually opposition from the general public on the whole system of reusing the wastewater. This is because they lack the relevant expertise and knowledge on how to actually apply the concept. Because of this, there are always high risks of been exposure to wastewater dangers due to reluctance of the people in reusing the wastewater. In other words, it is usually difficult to execute and implement any program involving wastewater reutilization without engaging the general public. Studies have revealed that one of the biggest problems that most companies receive in terms of wastewater recycling is lack of acceptance from the public. For instance, Canada citizens are usually uninformed about the importance of conserving and protecting groundwater. This is one of the major obstacles that various governmental agencies face when applying actions that are effective in addressing issues of water surety in the country. Therefore, prioritizing wastewater reuse in such a country is highly recommended in order to solve issues of water insecurity.

One best way of enhancing awareness of general public is through developing programs that will educate people on how to reuse wastewater and the importance that comes with it. Wastewater can actually acts as an alternative source of water supply in the cities. The reason why most people do not see the need of reusing reclaimed wastewater is because most of them fear the health dangers that are commonly associated with direct contact of this water. It is therefore the responsibility of the city authorities to clarify to the general public on the benefits and the level of treating wastewater. This will help in alleviating and managing various risks involved. For instance, acceptance of the general public on reuse of reclaimed wastewater can be enhanced through encouraging people to understand the quality of wastewater and the process involved in reusing it.

An assessment and management plan on risks associated to wastewater must be done before involving the general public in the process of recycling this water. There are several strategies that involve a risk assessment and management plan. These strategies include the following;

- Identification of potential hazards
- Assessing potential risk to exposure
- Dose-response assessment strategy
- Characterization of the risks
- Management of the risks
- Assessment on the ecological risks

All the bodies in charge of wastewater recycling process should undertake these strategies for the approval of the general public. It is imperative to also give feedback to the general public and try to assess all possible options of recycling the wastewater. Confidence of the general public can only be promoted through focussing on the past successes that have been achieved through recycling of wastewaters in the cities. Confidence on technologies and utilities used in reusing wastewater is also important in influencing the approval of the general public. The systems used in recycling the reclaimed wastewater must have integrated designs that provide the most appropriate safety guidelines. Other aspects that influence the success of a wastewater recycling program is having personnel that are well trained and conducting reliable monitoring of the program.

There are several constraints commonly associated with wastewater reuse programs. These include promoting public health and the required safety standards, adoption of feasible solutions, present and other emerging technologies in wastewater recycling programs, and limited wastewater recycling strategies details due to the confidentiality issues of the programs. Therefore, all strategies must be feasible, the program must present solutions in order to be accepted by general public, the solutions must reduce the current challenges experienced in city infrastructure, and the solution that must facilitate attainment of set goals and target.

ACCOUNTING FOR THE BENEFITS OF WATER REUSE

When the monetary value of all the benefits associated with a certain water reuse project are calculated and added up to a single dollar figure, then traditional engineering cost-benefit analysis can serve to compare the project to its alternatives. When all the benefits are assigned to a single agency or project sponsor, then such an analysis can also be used to determine the economic feasibility of the project. Unfortunately, neither of these situations normally prevails. In the first place, the benefits of water reuse should include watershed protection, local economic development, improvement of public health and other factors which are not readily quantified by traditional cost-benefit techniques. Second, the benefits are usually fragmented among a number of agencies and the public, and are not easily assigned. As a result, water reuse projects are often undervalued when compared to other projects, and significant opportunities for beneficial reuse are lost.

This section describes the many types of benefits derived from water recycling and suggests various approaches to quantifying them. It is not intended as a detailed economic analysis of water recycling in general, or of any one particular water reuse project. While several recent studies analyze the costs associated with water reuse, the benefits also need to be more clearly defined in order to properly evaluate the economic feasibility of recycling projects and accurately compare them with other alternatives.

Calculation and Distribution of Water Reuse Benefits

The economic value of water recycling projects is routinely underestimated during the planning process due to the non-monetary nature of many benefits. The traditional engineering economic analysis favors strict formulas and regimented calculation to arrive at an objective estimation of a project's worth. Such an approach is embraced by public agencies, lending institutions and regulatory bodies which share a sincere desire for precision, and which may be reluctant to make subjective judgments not required by policy. However, since certain benefits are not readily quantified (especially those with regional or global scope) this approach has the effect of setting the value of these "big picture" benefits equal to zero. To the resource economist, using analytical techniques commonly employed by that discipline, these benefits may be more valuable than the more narrowly focused benefits accounted for in rigorous detail through traditional cost-benefit calculations. Nevertheless, from the standpoint of the project planners, they do not exist.

The effect is compounded by the fact that local agencies already tend to view their responsibilities as service providers narrowly. As the directors and managers of these agencies see it, their fiduciary responsibility to their constituents (ratepayers or stockholders) requires them to provide their core service at the lowest possible cost. Despite the gradual trend towards viewing themselves as "trustees of public resources" water supply agencies still by and large define their mission in terms of supplying water, and wastewater agencies have as their goal the treatment and discharge of wastewater. With respect to water reuse, neither entity takes into account more than the current cost of the project and the benefit of the water produced or the diversion of effluent accomplished. Such a narrow perspective may avoid the introduction of subjective scalars, but it provides decision makers with a very limited tool with which to evaluate alternatives.

This situation will only change as the decisions based on such simplistic cost-benefit analysis are challenged, and more appropriate evaluations are offered in their place. This transition has already occurred in the arena of solid waste recycling. Three decades ago, the value of recycling solid waste was also in dispute. Advocates were viewed as extremists and their position was called impractical by opponents citing narrow, locally focused economic analyses. But with dwindling supplies of non-renewable resources and shrinking landfill space came the realization that solid waste recycling had benefits beyond short-term local economics, and efforts were made to identify and quantify those benefits. As a result, today the practice of solid waste recycling is widespread throughout the United States.

Even when all benefits are identified and evaluated, a water reuse project may not prove feasible so long as the benefits are distributed among entities which do not bear their cost. For example, a water reuse project may be sponsored by a wastewater agency to reduce the amount of effluent discharged into its receiving water ecosystem. Proposed as an effluent diversion alternative, the project should also provide a benefit as a new reliable, local source of water. This benefit, however, should not accrue to the wastewater agency, but to the local water utility which is responsible for meeting local water supply needs. As a result, the wastewater agency may not factor the water supply value of reuse into its planning process, even if the water agency is actively seeking new sources of water and is willing to serve recycled water to its customers. Or it may accept a value equal only to the current market price of water, without regard for the additional supply-related benefits of reliability or local control.

Similarly, if a water agency were to sponsor a water recycling project without the collaboration of the local wastewater entity, any benefits accruing to the sewerage agency would likely be either ignored or undervalued. Such disparity of interests is quite common in areas where the responsibility for water supply is distinctly separate from the authority to manage wastewater. The equally common result is that projects whose total value to the community exceeds their cost are not implemented and opportunities for reuse are lost.

This division of interests is universal, and stems from the traditional division of labor between those responsible for various aspects of water use, especially water supply and wastewater treatment. Even a centrally governed society like China maintains institutional barriers between water and wastewater agencies which have prevented implementation of water reuse projects. By contrast, in agencies where the entire water cycle is managed under the same administrative umbrella water reuse gains ready acceptance, because its diverse benefits are all appreciated. For example, the Irvine Ranch Water District, a California agency responsible for both water and wastewater utility management, has become a pioneer and leader in wastewater reuse. However, when responsibilities are not consolidated, the challenge is to bring the stakeholders together to identify the benefits of wastewater recycling, add them up and assign their costs fairly to all beneficiaries. The following section provides the outline of an approach to facilitate the identification and estimation of the benefits of water reuse.

Estimating the Value of Reclaimed Water

Not all values can be precisely or easily articulated in dollars and cents. Consider the following questions:

- What is the value of preventing endangered species from becoming extinct?
- What is the value of clean ocean beaches along the Atlantic and the Pacific Coasts of the United States?
- What is the value of business and industry retained in a community with a reliable, drought-proof supply of water?

These questions are not easy to answer, especially with traditional cost-benefit techniques applied by engineers in assessing the feasibility of wastewater recycling projects. But some method of calculation must be developed in order to estimate the real value of a project that:

- Prevents the extinction of endangered species in the receiving waters of a wastewater treatment plant;
- Redirects flow from an ocean outfall to local landscape and agricultural areas; and
- Provides a locally-produced, year-round supply of high quality water for industrial use.

In short, an environmentally conscientious project planner is responsible for providing a global perspective from which to assess alternative water, wastewater and water reuse projects. At a minimum, all benefits of water reuse projects should be described and discussed publicly, and appropriate quantification methods should then be used as needed to evaluate them properly.

VALUE OF RECLAIMED WATER AS A RELIABLE WATER SUPPLY

The most obvious benefit of water recycling is the recycled water supply itself. Supply is sometimes the only benefit identified, and is frequently set equal to the cost of existing water supply sources, either melded or un-melded. The melded cost of water yields a lower value, following the common utility practice of blending the cost of new and existing supplies to stabilize water rates. A somewhat higher value is obtained when cost-benefit studies use the unblended cost of the next best supply project as a proxy for the benefit of the recycled water alternative.

In contrast to the engineering approach, the economic approach tends to focus on the value of water, rather than its current sale price. For example, to estimate the agricultural benefits of recycled water, a resource economist would consider the market value of the crops grown, and the costs of all inputs to their cultivation, including water, fertilizer, land, management, etc. Normative linear and quadratic programming methods can then be used to isolate the value attributable to recycled water, which may be significantly higher than the current or future cost of water. This method will also indicate whether growers who receive the benefit of recycled water can afford to pay for the new water supply. A similar approach can be used to measure the value of industrial uses in the marketplace.

The benefit of recycled water to residential water users that do not have a market value can be determined as a function of their willingness to pay for water based on surveys of a representative sample of these users. Such surveys have been refined over the past twenty years in research and applied work. Questions are carefully crafted to avoid leading the respondent to overstate or understate values. As an added advantage, surveys can be designed to determine the value of reliability to potential customers of recycled water. This approach is also applicable to estimate other non-market values like environmental and recreational values, and is discussed in more detail in conjunction with those benefits. Reliability itself can be evaluated independently of the value of water. A recent study by the California Urban Water Agencies, based on a contingent valuation survey, estimated the value to residential customers of water supply reliability at $140 to $200 per acre foot. According to that study, "Monthly willingness to pay higher residential water bills to avoid shortages ranged from $11.60 to $16.90." Since the amount of recycled water produced increases with population growth, reuse projects can significantly increase the reliability of the local water supply. Other indices of the value of reliability include the cost of replacing landscaping which might be damaged during a curtailment or the surcharge on water rates for use during declared shortages. Including the value of reliability in a calculation of benefits enhances the feasibility of new water reuse projects.

VALUE OF RECLAIMED WATER AS EFFLUENT DISPOSAL

A typical wastewater management agency strives to meet its regulatory discharge requirements efficiently and economically. As these requirements become increasingly stringent, water reuse takes on an importance as a disposal strategy that may equal or exceed its worth as a new water supply. Given the limits of their traditional responsibility, wastewater agencies do not generally become recycled water purveyors without substantial economic or regulatory pressure. Regulatory agencies apply this pressure when they restrict or eliminate historic disposal options like river discharge, ocean outfall, and evaporation-percolation. Such was the case in the early 19th century during planning and construction of most cities in the world, when discharge requirements rendered disposal inordinately expensive.

The benefits of water reuse projects accruing to wastewater agencies are not typically counted when cost-benefit analyses are performed by other entities. In fact, unless the wastewater agency is the project sponsor, the benefits of avoided sewage treatment and disposal costs are rarely translated into monetary terms. Nevertheless, water recycling provides a number of significant benefits to wastewater agencies. In addition to the revenues from the sale of recycled water, typical benefits include;

- Reduction of effluent and preservation of discharge capacity
- Elimination of certain treatment processes to meet mass limits (e.g. heavy metal or nutrient removal)
- Delay or elimination of major sewers through construction of

satellite water reclamation plants

Although it is difficult to calculate savings in treatment, outfall capacity, and interceptor construction or distinguish them from the basic cost of meeting discharge requirements, once these savings are computed they can legitimately be assigned to the sponsor of a water reuse project. Like water supply benefit estimates, effluent disposal benefits are traditionally estimated based on the "next best" alternative disposal project. Resource economics uses the avoided cost model, too, but with a twist: it calculates the avoided cost of alternative effluent discharge, as well as the water supply value of the effluent. This cost may be indexed to anticipate future discharge requirements. In some cases, the wastewater agency realizes the economic value of recycled water to the region and partners with water purveyors to develop the resource.

VALUE OF RECLAIMED WATER FOR PUBLIC HEALTH PROTECTION

When wastewater treatment plant effluent is diverted from receiving water and reused, the quality of water in the receiving stream may be enhanced. In the case study of the Upper Occoquan Sanitary Authority (UOSA), improved receiving water quality was the goal of their indirect potable reuse project, which arose from a public health threat to the quality of water in the raw potable water reservoir. The UOSA water reuse project eliminated inadequately treated upstream discharges, consolidating them and providing higher levels of treatment and intensive monitoring. Project development of a safe potable water supply in this case also helped protect the public health against microbiological and chemical contamination.

Many communities living along rivers such as Mississippi, Ohio, and Colorado rivers in the United States rely on sources of raw water subject to upstream wastewater discharges. Reuse projects by any of the upstream communities would serve to protect public health, even if this benefit could not be widely touted for fear of disparaging the safety of current water supplies. This is even more often the cases in developing regions of the world where strict requirements for drinking water and receiving water quality are not enforced. In those places especially, water recycling projects can serve to improve both existing wastewater management practices and water supply conditions. Planners for water, wastewater, or public health sanitation have an opportunity in water reuse to accomplish several important societal objectives simultaneously.

The value of reducing infant mortality through a more hygienic water supply and wastewater management infrastructure, bridged with water recycling, is incalculable and priceless. Reduction of chronic enteric diseases and various epidemics can be accomplished through reuse of water in accordance with the minimum standards set by the World Health Organization. The economic value of reducing these public health threats is tremendous and must be rated highly even though monetary scales are inadequate for their expression.

Traditional cost benefit methods do not estimate public health benefits in dollar terms. However, resource economists are certainly capable of positing dollar values in these areas based on medical cost savings, decreased sick leave days and other indices of avoided cost. Even the benefit of reduced mortality can be estimated by means of actuarial calculations, extrapolated to correspond to the expected health improvement resulting from water reuse.

VALUE OF RECLAIMED WATER FOR ENVIRONMENTAL RESTORATION

The beneficial impacts of water recycling on the environment are both the most valuable and least attributable of all the benefits of water reuse. These direct environmental benefits include:

- Improved environmental conditions in upstream watersheds from which raw water supplies are extracted or exported
- Enhanced water quality in the receiving waters and improved habitat in the downstream watersheds through diversion of effluent
- Direct environmental restoration through stream flow augmentation, wetland restoration, urban park and agricultural development, and other "discretionary" urban use.

Monetization of environmental benefits is not usually attempted with any degree of precision in conventional engineering cost-benefit analyses. Nevertheless, the importance of these benefits is routinely stressed in support of water reuse projects, especially when conventional cost-benefit analysis determines them to be economically unfeasible. These benefits are also cited as accruing to the "general welfare" to justify project subsidies at various levels of government. It is through such subsidies (including grants, and low-interest loans) that the general public pays for widely distributed environmental benefits it receives from any water reuse project.

Again, while traditional engineering methods focus on the avoided cost of alternative environmental programs, the economic approach focuses on the value of the environmental improvements themselves as determined by various non-market valuation approaches. In one recent exercise, the value of the entire planet was estimated based on the value of environmental outputs like fish and fiber, and the cost of replicating environmental services such as atmospheric cleansing or oxygen generation. (The worth of the earth was estimated at $33 trillion.) Other methods whereby measurable impacts and public perceptions of environmental quality can be evaluated in monetary terms are described in greater detail below.

Upstream Watershed Protection

Throughout history, urban centers have developed close to sources of fresh water. Eventually, as population growth outpaced local resources, these initial water supplies became inadequate to support the population which depended upon them. Urban centers began importing drinking water from distant watersheds by means of monumental dams, canals, tunnels, pipelines, pump stations and reservoirs. For instance, in the Los Angeles Basin, about 60% of water supply is imported; certain areas of Southern California important 90% of the water they use. Even the relatively well-watered San Francisco Bay Area relies on imports to satisfy its water supply needs, from the Sierra Nevada Mountains, and from the ecologically fragile delta at the juncture of the San Joaquin and Sacramento rivers.

But upstream watershed areas cannot support unlimited water export without significant environmental impact. Water reclamation projects can reduce that impact, and even help forestall development of new supplies by damming canyons. The less water diverted from remote watersheds, the healthier the streams and habitats through which the water would otherwise flow. The value of upstream watershed protection gained legal recognition in the famous Mono Lake decision, which curtailed the rights of the City of Los Angeles to divert water from Mono Basin. Citing the importance of protecting upstream watersheds against excessive export of water, the court ordered reductions in water diversions until the level of the lake rose to a healthier level. The California Legislature voted to give the City of Los Angeles millions of dollars to develop water reuse projects to compensate for the loss of the Mono Basin water. The amount of money allocated by the State is one measure of the value of this broader societal benefit of water recycling.

Downstream Watershed Protection

As reflected in current discharge regulations, the minutest concentrations of certain pollutants in the treated effluent (e. g. heavy metals, pesticides etc.) can accumulate in the sediments of the receiving water environment and threaten the habitat of the local biota. In some cases, downstream habitat may be threatened by the sheer volume of the discharge, regardless of quality. For example, in the Silicon Valley area of Northern California, the discharge from the San Jose/Santa Clara Water Pollution Control Plant was found to threaten the habitat of two endangered species by converting nearby salt marsh to freshwater marsh. To prevent further conversion, the US EPA and the San Francisco Regional Water Quality Control Board required the City of San Jose and its partner agencies to implement a water recycling project that would divert 75,000 m^3/day (20-mgd) of freshwater effluent through reuse in irrigation of parks and golf courses, and a variety of industrial uses. Water recycling was found to be more expensive than the most practical alternative (construction of a deep water outfall). However, the value of protecting the downstream watershed, coupled with the value of the water supply generated by the project cumulatively outweighed its relatively high cost. Because of the importance placed on the protection of the two endangered species, both the state and the federal governments are assisting the City of San Jose with loans and grants to implement its water reuse projects.

Environmental Restoration

By providing high quality water in appropriate environments, recycled water projects have almost unlimited potential to protect, restore and create valuable habitat for wildlife; park land for urban amenity; and agricultural zones to promote sustainable open space and local food production. These benefits can all be evaluated using a variety of non-market valuation methods. Habitat improvement, species diversity, and even the health of a specific animal can be estimated and quantified for the sake of evaluating water reuse alternatives. The following types of values are all derived through representative sample surveys, interviews and other non-market valuation methods of natural resource economics:

Direct experiential or recreational values: These values are observed when a resource is directly experienced. Bird watching is one of these types of values.

Options on future use values: These values reflect the worth to the individual of being able to enjoy an environmental amenity in the future. An example would be the value to a person of being able to watch birds in the future as opposed to the direct experiential value, which is based solely on contemporary use value.

Existence value: These values are based on individual's appreciation of "just knowing" that environmental improvement, species and habitat health or survival of specific animals has been obtained. An example of this is the value of knowing that birds are healthy and alive, without actually watching them now or in the future.

Bequest value: These values are based on individual's value of improving the environment solely for the benefit of future generations. An example is the value of improved bird habitat to ensure the ability of future generations to be able to watch birds.

VALUE OF RECLAIMED WATER IN ENERGY CONSERVATION

Although it is often overlooked, the value of energy conservation attributable to water recycling is one of the more easily quantified benefits. Power costs are easily calculated and energy is itself a kind of uniform currency. Depending on the depth of analysis, this benefit can also be viewed either in terms of traditional replacement cost or avoided cost, or from the economic perspective of market or non-market value.

In Southern California, as in any other location where local water supplies are inadequate, water must be imported from hundreds of miles away. The costs of such imports vary greatly. For example, water from the Los Angeles Aqueduct travels several hundred miles from the Mono Basin/Owens Valley area. However, it flows by gravity and actually produces energy as it flows to the City of Los Angeles. On the other hand, the State Water Project supplies must be pumped from the Sacramento—San Joaquin Delta, over the Tehachapi Mountains and into Southern California at an estimated net energy cost of 3,000 kilowatt-hours per acre-foot.

In 1996, the volume of recycled water produced in Los Angeles County alone was 129,253 acre-feet. At a net cost of 3,000 kWh to pump each acre-foot of State Water Project water, a total of 388 million kWh of energy was conserved in 1996, equivalent to the annual output of a 44.3 megawatt power plant. At a cost of \$0.08/kWh, this equates to annual savings of \$31 million. These savings in electrical energy resulted in avoided consumption of over 200,000 barrels of oil. This reduction in oil consumption reduces the need to import oil via supertankers, which in turn reduces the potential for oil spills and the costly environmental damage that follows. In fact, the avoided consumption volume of oil in 1996 was almost as much as the oil spilled by the Exxon Valdez in 1989.

The conservation of fossil fuels and energy due to the use of recycled water also resulted in significant reductions in air pollutants in Los Angeles County. During 1996, 295 tons of nitrogen oxide, carbon monoxide, sulfur oxides, particulates and reactive organic gases were kept out of the atmosphere. Similarly, use of reclaimed water resulted in avoiding the production of over 340,000 tons of carbon dioxide, a greenhouse gas that is suspected of contributing to global warming. Health related economic benefits derived from the reduction of air pollutants include lower health care costs for respiratory ailments and decreased worker sick days, the magnitude of which are dependent on the proximity of the pollutant reductions to population centers. The economic benefits of decreased global warming are as important (and just as uncertain) as the very phenomenon of global warming itself. The most important benefit would be the reduced loss of coastal urban development to rising sea levels.

VALUE OF RECLAIMED WATER IN REGIONAL AND LOCAL ECONOMIC DEVELOPMENT

As an adjunct to its benefit as a reliable water supply, recycled water projects can significantly enhance the local and regional economy. Specific economic benefits include increased commercial and industrial development in the region because of the greater reliability of the local water supply, and creation of jobs directly due to project design and construction and indirectly due to location of new businesses in the area. These direct and indirect benefits can be estimated using input/output sectorial analysis. Input-output sectorial models provide the appropriate coefficients for income and employment based on the economic relationships between sectors in the local and regional economy. Common sectors include retail, construction, manufacturing and services, but additional sectors can be "disaggregated" from the community and be analyzed in detail as needed.

In evaluating a certain water reuse project, it is important for planners to examine the applicability and the quantified dollar value of each of the benefits attributable to the project or its alternatives. Obviously, not all projects benefit the environment to the same extent, nor do they all necessarily contribute to energy conservation. On the other hand, there may be some very specific and unique benefits peculiar to the particular project under consideration. Such site-specific benefits usually overshadow all others in importance and value. A summary tabulation of the benefits discussed above is presented below:

Table 7.1: Benefits associated with wastewater reuse

Benefit	**Applicability**	**Value ($/AF)**
Water supply	Very common	$300 - $1,100
Water supply reliability	Very common	$140 - $100
Effluent disposal	Very common	$200 - $2000
Public health	Situational	
Environmental benefits		
Upstream watersheds	Very common	

Downstream watershed	Common	$400-$800
Environmental restoration	Situational	
Energy conservation	Situational	0 to $240
Economic development	Situational	

Source: https://www.slideshare.net/pulkitshukl/recycle-and-reuse-of-wastewater

This listing of methodological approaches is not intended to be definitive or exhaustive. "Instead, the discussion was presented to provide the reader with a sense of what is possible in evaluation methods and concepts. Ultimately, the detailed focus of project evaluation should be based on a good understanding of major areas of impact, requirements for funding, and public acceptance. Sensitivity analyses that identify the key features that drive the selection of the preferred plan can also be used. The bottom line is that a broader evaluation of societal values should not be avoided out of ignorance or malice. It is possible to lose a valuable project to inadequate evaluation of the benefits received"

CHAPTER 8

HARVESTING AND STORAGE OF RAINWATER

DEFINITION OF RAINWATER HARVESTING

According to Rainwater Harvesting Handbook by African Development Bank (2008) water harvesting in its broadest sense can be defined as the "collection of runoff for its productive use". Runoff may be collected from roofs or ground surfaces. Additionally, runoffs can also be collected naturally for storage in the soil or aquifers and used for domestic, agricultural, livestock and other purposes. This technical chapter focuses on rain water harvesting for domestic use for households, schools, hospitals and other institutions. The collected rain water may also be used for kitchen gardens to supplement family food and income. In other places there are other rainwater harvesting activities that contribute to climate resilience such as different forms of terracing and collecting runoff from slopes for irrigation. These are very important and valuable technologies but are not further considered in this chapter.

BENEFITS OF RAINWATER HARVESTING

Collection of rainwater is of particular interest in areas where there are no suitable surface water sources and where ground water is too deep or too costly to reach. Rainwater harvesting is a relatively simple technology using mostly locally available building materials and skilled artisans. Rainwater has the added advantage to be relatively clean in contrast to ground and surface water which are exposed to numerous man made pollutants including fertilisers, pesticides and human waste by-products. There are several advantages and disadvantages associated with harvesting of rainwater. They are outlined as follows;

Table 8.1: Advantages and disadvantages of rainwater harvesting

ADVANTAGES	DISADVANTAGES
Relatively cheap materials can be used for construction of containers and collecting surfaces Construction methods are relatively straightforward Low maintenance costs and requirements Collected rainwater can be consumed without treatment, if a clean collecting surface has been used Provides a supply of safe water dose to homes, schools or clinics, encourages increased consumption, reduces the time women and children spend collecting water, reduces back strain or injuries from carrying heavy water containers	Supplies can be contaminated by bird ‘animal droppings on catchment surfaces and guttering structures unless they are cleanediflushed before use Poorly constructed water jars ‘containers can suffer from algal growth and invasion by insects, lizards and rodents. They can act as a breeding ground for disease vectors if they are not properly maintained

Source:*https://www.slideshare.net/rajashriamy1/rain-water-harvesting-26233867*

Rainwater harvesting helps in reducing the time required for collecting water. Statistically, it has been revealed that both women and children usually spend more than 30 minutes per day fetching water in rural areas and about 10-15 minutes in urban areas. Most of the water sources in rural areas are located within ½ km from the homesteads. Additionally, about 7.5% of women have to walk more than 3 km to access water for domestic use. Therefore, rainwater harvesting technique would really benefit many people especially in rural areas by serving the time they use to go and search for water long distances.

Rainwater harvesting is also beneficial in terms of reducing the cost incurred to buy water from municipals and other water sources. A survey done in one of the developing countries in Africa revealed that the average cost of a jerrican of fresh and clean water costs about a quarter dollar. Considering that one person uses about two jerricans of water per day, a household of three would expenses more than fifty U.S. dollars as expense for water. Therefore, rainwater collection may help in reducing this cost of water by half.

Erosion control is often mentioned as one of the positive impacts of rainwater harvesting. Capturing of the roof water results in less runoff water and reduces soil erosion on the steep slopes found in many places. However, it is not clear how big this impact of the rainwater harvesting and how this can be measured.

The impact of rainwater collection on the utility companies is not immediately clear. It has been shown that most utility companies are likely to lose revenue as rainwater harvesting techniques spreads because most people are likely to buy less water from their systems. There will result in a reduction in operating costs, especially less electricity consumption for the pumping stations. However, utility companies still need to ensure water supply to the population during the dry season when domestic rain water tanks run dry. The utility companies will therefore have to invest in the expensive infrastructure while losing some of its revenue.

Other benefits of rainwater collection include providing opportunities for different household to practise gardening and livestock rearing due to additional water supply. Health and hygienic benefits are also achievable due to additional water supply.

DETERMINING THE SIZE OF STORAGE RESERVOIR

Roof-top rainwater harvesting systems have basically four elements which are: (1) the roof of the building (which in most areas is usually made of iron sheets) (2) a conveyance system consisting of pipes and gutters), (3) the storage reservoir (such as masonry reservoir, plastic water tank, jerricansetc.) and (4) a delivery system consisting of a tap in the most simple applications or piping and pumps in larger dwellings.

From these four elements, the storage tank is the most costly part of the whole rain water harvesting system. Therefore it is important to select the size of reservoir that fits the requirements and financial capacity of the user. If the reservoir is too small, the users will have to face periods where they have to buy or collect water from other sources. A larger reservoir will increase the investments costs. There are several factors that should be considered when estimating the required size of the water reservoir for large public building. These factors include;

- Rainfall patterns; averages and variations of the seasons between years
- Catchment area in terms of size (m^2) and type of roof: flat versus slope, iron sheets, tiles or grass cover. For the common sheet covered roofs it is estimated that 85% of the rainwater falling on the roof is actually available.
- Use of filters which result into some losses
- Daily demand for water for sanitary uses, washing, cleaning, animals and possible homestead irrigation. Collected rainwater may also be used for drinking provided it is treated.
- Number of days that are acceptable to the user to be without water. This is an important factor and also depends on what reliable alternatives there are available close to the users such as utility company piped water, wells, springs etc.

All these factors should be brought together and used to calculate an optimal capacity of the reservoir. It is recommended that one should choose the smallest tank size that can meet an acceptable amount of the water demand. Any commercial/industrial rainwater harvesting system that can supply 70-100% of the required volume is generally considered to provide a good level of service. Large water reservoirs (build above or underground) can be constructed in sizes of up to 30 m^3. If more capacity is needed, additional tanks can be constructed. The plastic water tanks that have become popular over the last 10 years are a scalable system whereby tanks can be easily replaced or added at a later stage at relatively low extra costs.

Rainfall patterns: most places around the world have bi-modal rainfall patterns with the short dry seasons coming between the months of December to January and the long dry season from June to mid-September. The average annual rainfall may also vary from one place to another, especially between the drier lower parts and the higher altitude areas.

Daily Water demand: studies have revealed that daily household demand for domestic water in most developing countries stands at 25-40 litres/person/day as shown in the table below. The recommended storage capacity for a household of five persons which should last for about one month should be between 4 – 5 m^3.

Table 8.2: Daily water demand

Purpose/ person/day	Drinking	Cooking	Washing dishes	Personal hygiene	Washing clothes	Total
Amount (litres)	3-5	4-5	3-5	5-10	10-15	25-40

Source:https://www.epa.gov/watersense/how-we-use-water

These estimates above are only a guidance and need to be adjusted according to local conditions and the preferences of the users. For instance the estimates do not include the requirements for animals and many families around the world keep dairy cows at their homesteads which require about 20-40 litres of water per day per animal. There is also the increasing need for supplementary irrigation for homestead gardens, which are promoted by different governments as a means to improve nutrition and family income.

For homesteads: Studies have revealed that more than 45% of households around the world depend on utility companies' tapped water. About 30-40% households also use harvested rainwater for their daily domestic activities. Additionally, it has been revealed that more than 65% of households use jerricans as the main tool for rainwater storage. Other findings from different surveys conducted around the worlds revealed that;

- Drinking water counts for 8% of water usage, washing for 29% and cooking for 25%
- The harvested water is usually insufficient for the majority of people (about 75%) around the world. It lasts one month on average. Once depleted, many people use piped water or springs.
- Maintenance is not found to be a big problem
- About 20 – 30% of household around the world are usually affected by floods, excess runoff and erosion
- Major challenges involving water security include high costs, limited storage capacity and irregular rainfall

For institutions and businesses: About 35% of the institutions and businesses around the world use collected rainwater and the other 65% from utility companies. The main uses of the water include washing and cleaning (about 70%). Plastic and metal tanks water storage account for 60% and 10% respectively. Those made out of blocks or bricks account for 5%. Jerricans are also very popular with 20%. It has also been revealed that about 70% of the institutions and businesses use their stored water for a period less than one month.

COSTS OF RAINWATER HARVESTING TECHNOLOGIES

Investments Costs for Reservoirs

There are large variations in costs for different types of water storage reservoirs as illustrated in the table below. There are some differences between the two sources but these are not large. Actual prices also depend on location, availability of building material and craftsmen.

The reservoirs can be divided into those that are constructed in situ from blocks, stones, cement and reinforcement bars and pre-fabricated tanks from metals and plastic materials. Most of the pre-fabricated plastic and metallic tanks are usually manufactured mostly in cities by small industries. The water tanks made out of polyethylene (PE) are also used around the world and they come in different sizes depending on the suppliers and have become increasingly popular over the last years. For these pre-fabricated tanks, a simple foundation is recommended and these costs are extra. The following table illustrate different rainwater storage reservoirs and their estimated costs. The prices of PE water tanks need further verification against actual market prices.

Table 8.3: Types and units costs of different rainwater storage reservoirs

	DESCRIPTION	COSTS RANGE	OBSERVATIONS
In situ	Ferro cement tanks from 10 – 100 m³	\$0.5 – \$250/m³	Risk of cracking but can be repaired easily Known technology. Larger ones will require good artisans.
	Brick/masonry tanks from 5 – 30 m³	\$200 /m³	Long life Known technology, used for larger and smaller tanks Can be repaired easily
	Artisanal tank with plastic liner 6 m³	\$25/m³	Well established technology in some rural areas Local artisans can construct them Easy to repair

Pre-fabricated	Metal tanks 1- 10 m³	\$200 / m³	Locally manufacture Corrosion protection needed
	Fibre glass tanks 2- 75 m³	\$275/m³	Long lasting Easy to repair Limited number of suppliers
	Polyethylene tanks 0.5 – 10 m³	\$160 /m³	Widely available in urban areas Light and quick installation Requires foundation or other support construction Scalable, start with one and expand
	Jerricans 20 litres	\$150/m³	Widely available Easy start up with few jerricans and scaling up over time if funds are available Labour intensive (filling) Space needed inside house.

Source:*http://www.nzdl.org/gsdlmod?e=d-00000-00---off-0hdl--00-0----0-10-0---0---0direct-10---4-------0-1l--11-en-50---20-about---00-0-1-00-0--4----0-0-11-10-0utfZz-8-10&cl=CL2.12.5&d=HASH4d35564c531de1e8425e5e&gt=2*

Investing Costs of Guttering System

The other important requirement for a rain water harvesting system is the guttering system to collect the water from the roof and the pipes to bring it to the storage tank. These costs may vary from one place to another depending to the type of material used and the size of roof. The most common systems use metal or plastic gutters. In some cases people select the cheaper option of cutting and bending iron roofing sheets which works well for shorter distances. No separate costs are given for bends and connectors and it is assumed that these are included in the prices per unit of 1 meter shown below.

Table 8.4: Indicative Costs of Gutters

TYPE OF MATERIAL	COST U.S. $/M
Metal gutter	10-20
Plastic gutter	7.5-15
Iron sheet (roofing material)	5-10

Source:*https://www.slideshare.net/CTAseminar2010/rainwater-storage-options-mechanisms*

For a small house of 120 m^2 floor space about 30 m length of gutter will be required which would be U.S. $ 1-300 for a complete system, using the unit prices above.

Investment on Other Accessories

Other accessories might be installed depending on the needs and wishes of the users. A water diversion valve can be installed to reduce the contamination of the initial water flows coming from the roofs. Water filters can help remove sand and other dirt before the water flows into the buildings. While many households will bring the water into the house using buckets and jerricans, higher income groups and public building may want to install an electric pump and overhead tanks. Costs of the required plumbing and electrical materials are not further included in this technical note but can be obtained from specialised contractors and hardware shops.

RAINWATER HARVESTING FOR HOUSEHOLDS

Water harvesting practices in the form of plastic cistern has always remained the centre of main activity and struggle of all the people around the world. These structures have been serving the human and livestock population besides very precisely supporting the system of irrigation. Rainwater harvesting systems play a crucial role in most places around the world, especially in arid belt, where in most of the areas the ground water is saline and unsuitable for drinking, and rainfall is erratic and unpredictable. Water harvesting systems are efficient, easy operative and cheap in the long run in many ways. Among the rainwater harvesting systems used by people, the cistern can be the answer for reliable and safe drinking water source. Its structural important features, optimize capacity, efficient catchment area characteristics, size required and after care are discussed here.

Figure 8.1: Underground rainwater storage cistern.

Source:*http://www.yourhome.gov.au/water/rainwater*

Structural Features

A cistern is a covered underground facility, generally of masonry (stone or brick) concrete, or plastic for collection and storage of surface runoff. The provision of cisterns near religious centers and in village for community use is an ancient practice since 1607 A.D. However, nowadays it is preferred to construct it at each village on the individual ownership basis. It is mainly because in case of community cistern, the dominant group gets individual supply of water to their homes which poor do not receive. Moreover, an access to water supply might be monopolized. Generally, community structures failed through poor workmanship and after care.

Figure 8.2: Artisanal water reservoir under construction and completed with roof.

Source:*http://dbsinfrastructure.com/water-infrastructure-water-storage-tanks-ponds.php*

Design Components

Rainfall probability: Annual rainfall, distribution, amount, and intensity plays an important role in designing of catchment area and structure. Most of the times the structures designed on the basis of average rainfall do not serve the purpose because of frequent droughts in the arid region.

In general, the water supply structures are either over designed or under designed. Considering the peculiar rainfall conditions in the arid and semi-arid regions, it may be worthwhile to predict probability of dependable rainfall of that particular zone where the structure is being proposed. Studies indicate that the dependable annual rainfall occurring at 60 % probability worked out through the Log Pearson Type III distribution meets the criteria for suitably designing the catchment area for the cistern. By considering this probability of rainfall and catchment area needed, the tank meets the water requirement throughout the year.

Design of cistern: The circular cross section is the most economical form of the cisterns. It has also been found to be more stable as the pressure exerted by the water is uniform radial pressure in all directions in the diametrical plane at right angle to the curved surface. The successful installation of tank depends on the selection of site, particularly the catchment characteristics such as size, shape, topography, soil type, vegetation etc. In case of natural catchment, the size of the catchment artificially prepared should produce adequate runoff to meet the storage requirement of the tank and at the same time it should cause minimum soil loss. The catchment area could be worked out based on the following equation;

$$Ac \times Pd \times C = m \times d \times Ao$$

Where

$Ac = Catchment\ area$

$Pd = Dependable\ rainfall$

$C = Runoff\ coefficient\left(found\ to\ be\ 0.3\right)$

$m = Storage\ coefficienct\ taken\ as\ unity\ for\ the\ tank$

$d = Depth\ of\ the\ tank\ below\ G.L\left(3.0M\right)$

$Ao = Cross\ sectional\ are\ of\ the\ tank$

The catchment area required for an ideal capacity tank is worked out to be 330 m^2

After selection of the site and size of the catchment, in the case of natural catchment, it is imperative to carry out some minor treatment for augmenting the runoff. These treatments may be such as construction of earthen diversion bund, removal of bushes, increasing the land slope, compacting soil etc. However, in case of artificial catchment, it should be round in shape, giving proper slope of 2.5- 3.0 % towards the tank with zero slope at inlet. After forming the shape of the catchment, it should be rammed and spreading 15 cm thick layer of calcium carbonate nodules (murrum) and again ramming. In case calcium carbonate nodules are not available, a layer of silt clay can be spread and rammed properly. Both the treatments on the artificial catchments will result in reduction in the intake rate of the soil.

In both the cases of the treatments to the catchment, the maintenance is required every year. Moreover, during runoff, the treated surface is worn out and the materials are carried out as suspended load and deposited in the tank. This deteriorates the quality of the stored water. To avoid these, a courtyard or open area of the hamlet is treated with cement plaster or cement tiles laid on the bare soil. In this case, the additional water is collected of better quality from the treated area. The coefficient of runoff becomes 0.9 instead of 0.3 due to pucca floor. Hence, maximum water is harvested from small area for storage. The runoff water is transferred through a pipe network to the tank. The catchment requirement and the cost estimates will be presented later on.

Structural components of cistern:

- Catchment – It should be impermeable and smooth to obtain improved hydraulic efficiency
- Silt collecting pit – It is provided to arrest the sediment load before the inlet point
- Cemented apron – This is one of the main structural components provided to control seepage and also to prevent caving from sides. Suitable cement concrete 1: 3: 6 is being used for construction
- Catch pit – This is made in the form of depression in the foundation slab of cement concrete 1: 3: 6 to facilitate drawing minimum available water from the cistern
- Top cover – It is required to control evaporative losses and to prevent pollution of water stored by foreign materials. An opening at top with G.I. cover to facilitates withdrawal of water
- Inlets – There should be three inlets to collect the runoff, and

will also prevent floating debris from entering into the cistern. Moreover, by providing these, there will be free access of air in the cistern which will prevent the bad odor of stored water. A suitable mesh with iron bars in the angle iron frame is to be provided in the inlets

- Outlets – There should be one outlet for letting out the extra water entering in to the cistern
- Guide walls – These are provided upstream of outlet for arresting all the available runoff from the catchment.
- Withdrawal arrangements – A stand with pulley at opening is provided for withdrawal of water with the help of rope and bucket.
- Fencing of catchment area – Fencing of the catchment area is required to restricted animal entry

Constructional details: The construction materials used are, all locally available such as stones or bricks, cement, iron sheet, etc. After selection of site and carrying out desired earth work in excavation; 3.25 m depth and 3.0 m diameter, cement concrete foundation 1: 3: 6 with 0.25 m thick should be laid. Entire construction of stone masonry in 1: 6 cement mortars should be done with special care to ensure against any crack and leakage from floor and walls of the structure. This is achieved by providing cement concrete flooring and at least 1- 1.5 cm thick cement plaster on walls, incorporated with water proofing compounds or with 1 % soap solution in water (soap solution forms insoluble fillers on reaction with cement). Proper cementing of joints in the super structure should also be ensured, especially when further construction on a half-finished work is resumed after the lapse of a few days. The cistern is covered with stone slab roofing with a suitable opening on the top for withdrawing the water.

The inlets and outlet in the cistern should be placed 3.0 cm below the ground level that is the point where the artificial catchment ends with cistern wall. All the inlets and outlet should be made of angle iron frame with vertical iron bars spaced at 5 to 7 cm apart and expanded metal mesh to prevent entry of floating debris, birds and reptiles, etc. which attempt to approach the water, usually drop inside the cistern and spoil the water. The inward sloping cement concrete apron around the cistern with about

3 % slope helps in collection of even meager runoff. Moreover, this will also prevent the entry of runoff water along the outer face of cistern wall. It will also reduce the possible damage to the structure during high runoff rate of heavy rainfall event. The masonry check wall nearer to the outlet (both the sides) of about 50 cm. High constructed across the apron will divert the water into the cistern. Thorn fencing may be provided around the catchment area to minimize pollution of the stored water with livestock wastes and other farm wastes.

Maintenance and after care: At least once in a year cleaning of cistern, ramming of its catchment, and painting of the inlets and the outlet is must for keeping it in good working condition. Also the silting of the cistern is necessary, which may be taken up before the onset of the monsoon. Periodically use of oxidizing agents like potassium permanganate will prevent the growth of microscopic organisms, and development of bad taste, odor and color of the water. Alum helps in settling the dirt in suspension. At least a few centimeters of water column should always be maintained in the cistern to avoid development of cracks etc. if the cistern is well constructed and all structural requirement considered, it is expected that this cistern will last at least for 30 years

It is unpleasant truth that water availability in most arid regions is limited and inadequate, even after a hypothetical consideration that all utilizable water resources are developed. Further with increasing population and modernization, the scarcity of water will go on increasing and per capita water availability will be decreasing. Moreover, in less than normal rainfall years the situation is much more critical. Nonetheless, the importance of the old tradition of water conservation / harvesting was reflected in the seventeenths century when the structures were developed. Particularly in the rural areas, the water harvesting structures have served as the best solution in mitigating the drinking water problem. But the fact remains these water harvesting systems should be revived and encouraged at government level.

Water Demand and Consumption

Estimating the demand of drinking water and water needed for cooking is the prime important for designing any rainwater harvesting system. Studies have revealed that an average daily water consumption which includes animal consumption is about 30-60 liters / day in most cluster villages. In reality it is very difficult to separate out the water consumption by human and livestock. It may be only approximation. However, as per the WHO recommendations, minimum water quantity standard is 20 liters per day for human survival.

Basically, the runoff water stored in the cistern is meant for human drinking and cooking only. Animals are not supposed to be provided water from this structure. If the water is used by both, then the size of the cistern will be large or it is to be refilled by transported water which may be uneconomical to the poor rural population. The size of cistern, for a single family, of capacity of 21,000 liters for drinking only is sufficient. In case the provision for cooking and animals drinking water is required then the requirement can be calculated using the following figures;

Table 8.5: Livestock water consumption

Type of animal	**Age (year)**	**Water consumption (liters per day)**
Heifer	1	10-25
Mature cows	-	30-70
Sheep and goat	More than 1	2-5
Donkey	-	20-45
Camel	-	25-35
Buffalo	More than 1	30-70
Horses	-	25-50

Source: *http://www.yourarticlelibrary.com/dairy-farm-management/water-supply-to-dairy-animals-on-farm/36207/*

The requirements of drinking water for different type of animals can be passed by the above table and for human consumption and for cooking, 20 liters per capita per day could be considered according to WHO recommendations. Based on these data, at initial stage approximately the capacity of the cistern can be decided for each family. However, by collecting data on actual consumption of water per capita per day and by animals on average basis, it will be possible to standardize the construction size of cistern for given number of family members and animals along with extra allowance. The estimation of consumption by human could also be done by the following method.

The maximum yield possible from rain water catchment system is directly proportional to the catchment area, its runoff coefficient (i.e. the ratio of harvestable rainfall to total rainfall) and rainfall probability. Thus, for a full hydrological year, the main daily yield is given by the equation:

$$Y = \frac{fi \times Ri}{365}$$

Where

$$Y = mean\,daily\,catchment\ yield, liters\ / m^2$$

$$fi = runoff\ coefficient$$

$$Ri = probable\,rainfall\,in\,millimeters$$

The runoff coefficient depends on the catchment treatment material and age of the catchment surface. Theoretically, the consumption can be set at constant level of Y liters / m^2 /day if adequate storage is available for dry season requirement. This is based on the assumption that evaporation and other losses from storage are negligible. Per capita consumption from catchment is calculated by:

$$C = \frac{Y}{N}$$

Where

$$C = per\ capita\ consumption, liters\ / m^2 / day$$

$$N = number\ of\ persons\ served\ by\ the\ catchment$$

$$Y = mean\,daily\,catchment\ yield, liters\ / m^2$$

Example:

Six persons are to be served by cistern stored water, having treated catchment area of 500 square meters (f = 0.2). Rainfall probability is 212 mm.

$$\textit{Mean yield} \quad Y = \frac{212 x 0.2}{365} = 0.116 \text{ liters} / m^2 / \text{ day}$$

$$\textit{Per capita consumption} \quad C = \frac{Y}{N}$$

$$= \frac{0.116 \text{ x}500}{6} = 9.66 \text{ liters per day}$$

Therefore for six member family the annual requirement will be around 3,500 liters.

Size and Runoff Efficiency of Catchment Areas

Due to the high variability of hydro meteorological phenomena it is not possible to harness the available rainfall to optimum level, unless feasible measures are taken. For generating reliable runoff, a water harvesting catchment needs to be impermeable and smooth. The perfect sealing of the soil surface is little bit expensive; therefore, some infiltration losses must be accepted. In many cases, ground forming can increase the runoff yield by improving the hydraulic efficiency of the catchment and thus reduce the time available for infiltration to take place.

Figure 8.3: Typical pond/cistern for storing run off water.

Source:*https://www.peakprosperity.com/wsidblog/91878/how-build-earthen-pond-and-dam-wall*

Normally the poor rural population cannot afford the expenses on the treatment of catchment surface for generating the runoff. But, they can be advised to either go for enlargement of the existing catchment with same type of murrum treatment so as to harvest the runoff up to the capacity of the cistern or use cistern water for human drinking and cooking only. As usual they must take their animals either to the other water point such as village tank, or nearby water supply scheme. In case the size of the existing catchment cannot be altered then the surface of the catchment should made smooth which may have 0.8 or 0.9 runoff coefficient. Several experiments on use and efficiency of different materials have been carried out in different countries. Such treatment materials are; sprayable asphalted compounds, plastic and metal films bounded to the soil, soil compaction and dispersion field fabricated asphalt, fiberglass membrane, sodium salts, gravel covered plastic membranes and melted wax, bentonite and few chemical treatments. It can be cemented surface also. The life of continue cemented surface in open area is less durable and cracks develop on the top of it. To avoid this problem, cemented tiles of 30 x 30 cm on bare sandy soil can be provided without cement joints. Such surface will improve efficiency of the catchment area in runoff generation to the extent up to 90 %.

RAINWATER HARVESTING FOR PUBLIC BUILDINGS

Public buildings are usually ranked in different categories depending on size. Most of them are used to collect rainwater especially in developing countries such as Africa and some Asian countries.

When designing rainwater harvesting system for public buildings, there are certain elements that one should consider. These include rainfall data, roof areas, losses, daily demand for water, construction and operating costs and other factors. The table below provides an overview of the three categories of public buildings and their main characteristics based on the requirements for rainwater harvesting.

Table 8.6: Illustrating different forms of rainwater harvesting for public buildings

TYPE AND MAIN CHARACTERISTICS	LAYOUT
High rise buildings Roof area 500 – 1,000 m^2 Flat roofs, cement Underground reservoirs Pump to bring water to header tank on the roof Estimate average costs; U.S. $60,000	
Large buildings Roof area 1,200 to 6,000 m^2 Iron sheet roofs Schools, health centres, markets, district offices Large storage; under of above ground Water pumped back to roof tank or direct throuhg pipe network Estimates average costs: U.S. $80,000	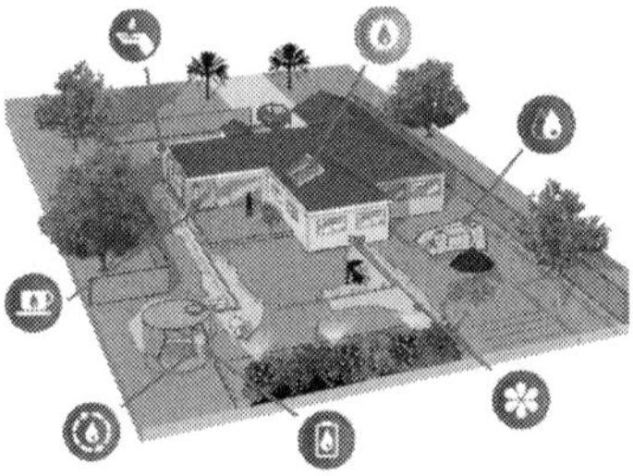
Small buildings Roof area 100 – 500 m^2 Iron sheet roofs Above ground tanks: cement or plastic Estimated average costs: U.S. $12,000	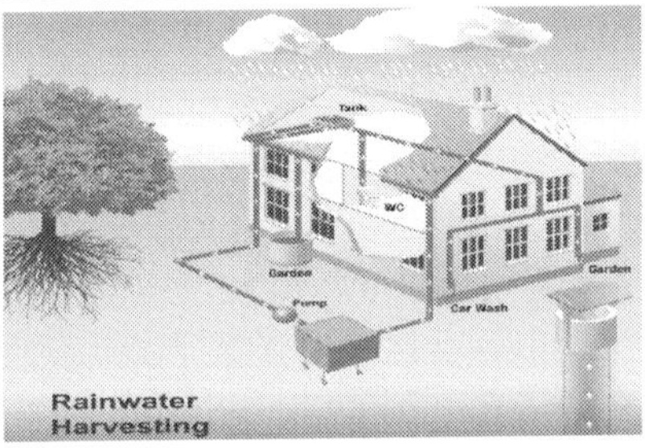

Source:*https://www.slideshare.net/sagarian/rain-water-harvesting-complete*

RAINWATER HARVESTING FOR IRRIGATION PURPOSES

Agriculture is the biggest user of water in most parts of the world, especially in developing countries accounting for up to 70% of the water demand and is expected to increase to 80% by 2020. People especially those from tropical and dry regions are always encouraged to increase irrigation to boost agricultural production. For instance, a recent study estimates that Rwanda has a potential of 589,000 hectare of irrigated land and the government targets to increase irrigated area from 18,000 ha in 2010 to 100,000 ha by 2020. This six-fold growth in irrigated agriculture will result in more pressure on the country's water resources.

Rain water harvesting for small scale irrigation has been tested and promoted in different African countries for many years. For instance the work done by the Kenyan government under Galana-Kulalu irrigation scheme using modern irrigation technologies to identify best opportunities for rainwater harvesting systems for agricultural purposes is a clear indication that a country can really benefit from harvested rainwater. Different governments also work closely with local communities as water harvesting is more than a matter of constructing ponds, dams or tanks. It requires involvement of the local communities in setting up and maintaining the system and share the water sustainably and equitably. Earthen ponds of different sizes using plastic lining have been constructed in several parts of Africa too such as in Rwanda for collecting rainwater from nearby slopes, roads etc. Most of the constructed rainwater harvesting ponds are usually equipped with a rope and washer pump to support water distribution.

Figure 8.4: Pond with rope and washer pump.

Source:*http://africawindmill.blogspot.co.ke/2011/11/awp-manualpedal-water-pump.html*

Currently, the total number of irrigation ponds in Rwanda is given as 1376 out of which 735 are working, 171 requiring repair, 350 planned/implemented and 120 ponds are under construction. These are in 16 relative drier parts of the country. The ponds are of the storage capacity of 120, 250, and 480 m^3 irrigating 0.25ha, 0.5ha and 1ha respectively. Assuming an average of 0.5ha, the total area that can be irrigated from these surface ponds is estimated as 688ha. The investments costs of these water reservoirs are given as Rwf 1 m for 120 m^3, Rwf 1.5 m for 250m^3 and Rwf 2.2 m for 480 m^3. More information on the use of ponds and small dams can be retrieved from various websites and governmental database and in particular from the Irrigation and Mechanisation ministry.

CONCLUSION

In summary, the concept of rainwater harvesting is not new in most parts of the world especially in arid and semi-arid regions where acute problem of drinking water exist. It has been a traditional practice since last four centuries. Runoff is being harvested from natural catchments, in the form of dugout ponds or impoundage reservoirs and from roof tops or artificial catchments in the form of cisterns. However, no sizable work on water harvesting from treated catchments have been initiated in most regions. The limitation to use of treated catchments has been the cost per unit of water produced. Nevertheless, there is importance of the treating the catchments so as to render the soil surface impermeable, thereby, eliminating infiltration and generating more runoff for water harvesting structures. Studies have revealed that water harvesting efficiency up to 60 % can be achieved at affordable cost. On murrum surface layer, polymers may be spread to augment the surface runoff. Second, alternative could be the use of cement tiles in the catchment area (Coefficient 0.9). Though, the initial cost will increase on runoff generation component but there will be assured availability of water in the cistern.

CHAPTER 9

SOLUTIONS TO WATER SECURITY CHALLENGES IN THE WORLD

In March 2000, water security was one of the many concepts that were highlighted in the "Framework for Action" in The Hague Ministerial Declaration. This declaration was outlined as a mean for curbing various challenges experienced by people from different parts of the world due to water crisis in the 21st century. For a long time it has been mentioned that people are able to take advantages of various opportunities presented by water resources if they are able to adequately access the water. Water security concept entails developing viable mechanisms that help all the poor people have a safe and supportable access to water resource. There must be a strong association between participatory functions and governance to help in dictating this access. Implementation strategies to solve problems of water security can be classified into four main categories as outline below.

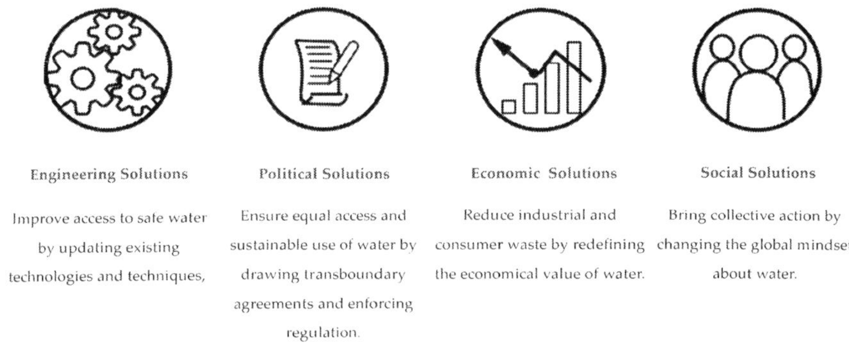

Figure 9.1: Categories of solutions to global water security problems.

Source:*http://12.000.scripts.mit.edu/mission2017/solutions/*

These four categories collectively represent a framework that can be used by all people, non-governmental organizations, and different nations to ensure that there is safe and reliable access of water for all.

ENGINEERING SOLUTIONS

Water security can be summarised into three important issues. These include access to safe and fresh water, pollution of water, and efficiency in water utilization. These issues can only be addressed through adoption of proper infrastructures. Application of engineering solutions towards curbing issues on water security involves identification of appropriate technologies that can help in solving various economic, social, and regional hurdles.

Access to safe and Fresh Water

One of the crucial elements of water security is provision of fresh and safe water for every person. However, most places face different challenges due to inconsistency in water supply considering the fact that demand is usually constant. For this reason therefore, one of the best way to ensure that there is consistent water supply for all is through stabilizing water supplies through the use of various infrastructure (such as pipeline and dams). Notwithstanding different efforts that are put by different nations in ensuring that there is adequate water for all, about one third of total world's population still do not have access to safe and clean water.

Access to safe water can be limited due to either physical or economical water scarcity. In other words, economical water scarcity entails existence of safe and clean water, but unaffordable for all (either based on time spent to obtain it or its high cost). One of the key contributing factors to economical water scarcity is poor and unstable water supplies and distribution systems. On the other hand, physical water scarcity means lack of adequate and clean water resources in a region. There are different factors that contribute to physical water scarcity in a place. These include overuse of the available water, insufficient precipitation, and lack of enough groundwater reserves in a place. Some of the places faced with physical water scarcity include North Africa, South East Asia, and central Australia regions.

These two maps illustrated below provide occurrence of natural water resources in comparison with their economic availability. Comparing these two maps, several disparities in terms of availability of water attributed to poorly established infrastructures can be observed.

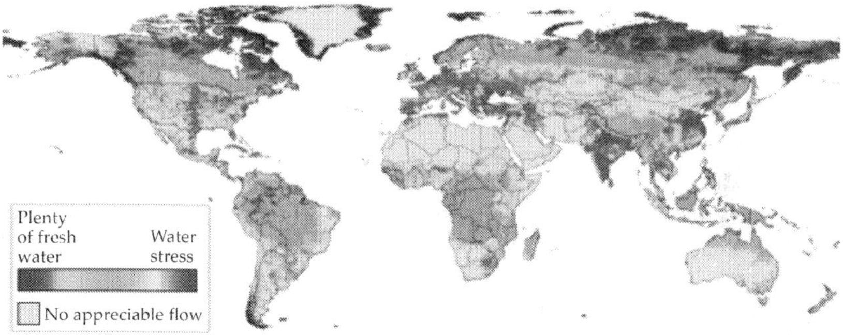

Figure 9.2: Natural occurrence of water resources.

Source:*http://12.000.scripts.mit.edu/mission2017/solutions/engineering-solutions/*

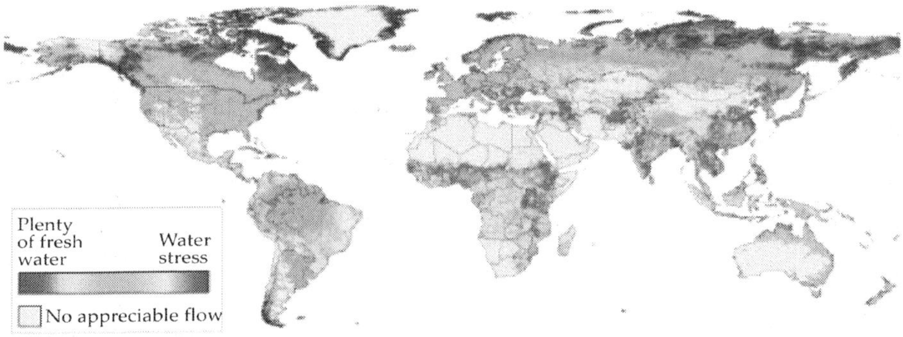

Figure 9.3: Managed water access with poorly instituted infrastructure that decrease availability of natural water.

Source:*http://12.000.scripts.mit.edu/mission2017/solutions/engineering-solutions/*

There are several integrated solutions that can be adopted in order to address the issue of poor water access in different regions of the world. One of the key steps towards enhancing water security globally is improving access to water in various regions with water scarcity and that are highly populated. Looking at the maps given above, it can be deduced that Australia, Asian and African countries are the most affected in terms of water scarcity. It has been revealed that this problem is usually accentuated by climate changes in most of the regions. To address this problem, solutions towards water scarcity can be outlined in to two causes as mentioned above (that is economic and physical water scarcity).

Economically, most of the places with severe water scarcity such as central Africa and south Asia lack sufficient precipitations, water resources and various mechanisms used in distribution of water. In most of these regions, water scarcity can be solved through implementation of proper mechanisms for harvesting rainwater both in the urban and rural areas. This method is beneficial in one way or another because it is less costly and it can also be adopted at any scale (that is at municipal level or personal level). To have a long lasting solution to economic water scarcity especially in the urban areas, it is important to centralize various water supply and distribution systems. This will helps in easing various activities done to improve water supply such as the use of proper hydraulic pressure, material replacement, and the use of better maintenance through regulation practices.

There are different regions faced with physical water scarcity that cannot solely depend on harvesting of rainwater due to insufficient precipitations. Some of these regions include North African countries, central Asian countries, and Australia. However, these places require creation of new water sources and maximizing the present water resources. One of the first and most important steps towards creating self-sustainability in water supply in these areas is constructing plants for recycling water. This mechanism is one of the cheapest solutions to water scarcity and it helps in doubling water supplies. Additionally, the wastewater that is not recyclable can be used in replenishing various aquifers that have been depleted. This wastewater is purified naturally over a period of time. Another method to improve on water supply in regions with physical water scarcity includes developing infrastructures for desalinating the saline aquifers and seawaters.

Figure 9.4: Desalination of seawater.

Source:*https://constructionreviewonline.com/2017/02/desalination-using-reverse-osmosis-system/*

Enhancing access to safe and clean water also includes developing mechanisms to ensure that there is adequate water for future use too. Those places with unsustainable groundwater such as in China, United States, and Central Asia should look for other alternative freshwater sources. For instance, municipal should developed proper mechanisms for harvesting and capturing rainwater that could otherwise go to waste. Issues on climate changes on water scarcity should also be addressed amicably. Some of the proposed strategies towards dealing with water scarcity issues include;

- Creating fixtures that are permanent to help in stabilizing distribution and access of safe and clean water
- Decentralizing all systems for rainwater harvesting and distribution to people
- Stabilizing precipitation through reduction of carbon dioxide emission
- Dams and reservoirs decommissioning that are cost-effective
- New projects vetoing in order to reduce overreliance on dams
- Creating plants for water treatment and desalination
- Recharging aquifers that have been depleted using grey water obtained from different treatment plants

Water Pollution

Studies have revealed that for every ten people, one usually has limited access to safe and clean water. A huge number of people also die each year because of using water that is unsafe for their health. Water pollution is mainly caused by wastes from human activities. Statistically, about two million tonnes of untreated agricultural, sewage, and other industrial waste materials are usually dumped into waterways in the world. In least developed and developing countries, more than 75% of wastes from industries are usually disposed directly into the lakes, seas and rivers.

Access to fresh and clean water is one of the key human rights that many people are not aware of. Considering the rapid world population growth rate that is currently been experienced, there is the need to develop ways of curbing different water pollutions. This will ensure that right to clean water access is maintained. There are certain parameters pertaining water pollution that should be addressed. These include lack of awareness and inefficient governmental regulations on danger associated with water pollution.

Figure 9.5: Industrial freshwater pollution.

Source: *http://friendlylivinggroupblog.blogspot.co.ke/2015/04/water-pollution-contamination-of-lakes_10.html*

There are several integrated solutions that can be used in addressing key sectors that cause the highest pollution of freshwater globally. These sectors include agriculture, industries, and sewage as mentioned here above. Some of the worse polluted water sources are located in Africa and south Asia. The main source of water pollution in these regions is untreated sewage effluents. To help in curbing this menace, the responsible authorities should ensure that there are enough latrines in the rural areas and proper sewage system constructed in the urban areas. It is also important to educate people on the need to safeguard water sources in order to prevent pollution and control various water borne diseases.

Some of the cities that are poorly regulated (such as Guangdong in China) face water pollution mainly from industrial discharges rather than from domestic effluents. There is the need to form key environmental regulatory agencies in these cities that will help in addressing water pollution through enactment and enforcement of different regulations. Some of the key regulatory methods that these agencies can use include trade and cap. In other words, the money that is obtained from different caps and trades should be utilized in funding projects developed for purifying water that has already been contaminated from different sources.

It is also important to enlighten different industrial companies on the need to invest on technologies that are less likely to cause pollution on water sources for instance, filtration and recycling of industrial effluents.

Figure 9.6: Key causes of industrial water pollution.

Source:*http://helpsavenature.com/industrial-water-pollution*

Some of the region where agriculture is the key pillar on the GDP for a nation, it can be one of the biggest pollutant on water sources. According to estimates given by EPA, it has been revealed that agricultural runoff in some of the developed countries such as United States accounts for more than half of all river impairments and pollutions. Most of the runoffs are attributed to insufficient communication and awareness. One of the key strategies that can be adopted at community level is developing programs that help in enlightening people on how to practise better farming practices. For instance, educating people on how to shift from various farming practices that cause intense pollution on water sources (such as chemical supplementation and monocropping). Some of the best and clean farming alternatives include organic and crop rotation farming. In summary, some of the key proposed strategies and plans for control water pollutions include;

- Treating and regulating wastes from industries through the use of trade and cap and other water treatment facilities
- Improving facilities for treating sewage through both decentralization and centralization
- Improving irrigation and farming practices
- Implementing proper technologies to help in reducing problems with both agricultural and industrial waste management

Water Efficiency

The current global situation on water usage is unsustainable. Therefore, there is the need to reduce this utilization in order to improve and secure sufficient water access for the future generations. Studies have revealed that the rate at which water is been used in various developed countries such as Europe and United States is beyond its replenishment rate. According to the figure given below, it can be seen that energy and agriculture are the two key sectors where water is heavy utilized. This figure illustrates various ways in which water is withdrawn (water taken from various water systems) and consumed (water that is not accessible for human use) by various sectors and the continent.

This figure outlines key regions with high volumes of water withdrawals per continent. These volumes are based on the differences that exist in terms of population and water consumption ratios to water withdrawal. In other words, water withdrawal and consumption ratios can help in evaluating efficiencies in water usage between regions.

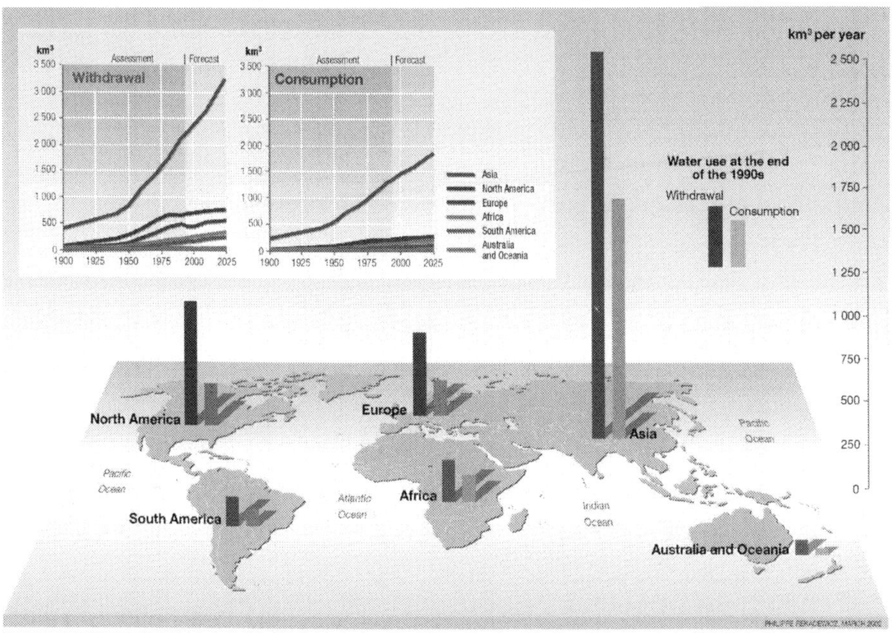

Source: Igor A. Shiklomanov, State Hydrological Institute (SHI, St. Petersburg) and United Nations Educational,Scientific and Cultural Organisation (UNESCO, Paris), 1999; *World Resources 2000-2001. People and Ecosystems: The Fraying Web of Life*, World Resources Institute (WRI), Washington DC, 2000; Paul Harrison and Fred Pearce, *AAAS Atlas of Population 2001*, American Association for the Advancement of Science, University of California Press, Berkeley.

Figure 9.7: Global water withdrawal and consumption.

Source:*http://www.colinmayfield.com/waterhealth/course1/content/discussion/wh001mo001c001d160Importanceofwater.html*

There have been several integrated solutions proposed to help in dealing with water scarcity in most parts of the world. One area that has received more emphasis is agricultural water usage. Studies have revealed that more than 75% of freshwater consumption in the world goes to agricultural activities. Some of the main areas that account the largest agricultural water usage include South America, Australia, and Africa. This also translates to highest freshwater wastage in these areas. Most of the water in these areas is lost through evaporation and runoffs.

One of the best ways to address and curb these problems is enhancing ways used in flood water irrigation (such lining canals with clay or concrete) and also by employing surge flooding to alleviate water runoffs. Most of these irrigation techniques are usually easy to apply and cheap too. Studies have shown that use of proper irrigation techniques can help in reducing the amount of freshwater consumed through agricultural activities by more than 25%.

Apart from agricultural activities, production of energy is another key area that accounts for the highest water consumption in the world. Statistically, it has been shown that water consumption in various energy production sectors in the world is increasing at an alarming rate (for instance, United States energy sector accounts for more than 27% of water consumption in the world). One way that can be used in reducing water wastage through energy production is replacing various "water hungry" fuels such as coal and biofuels with other fuel sources that are more efficient. In summary, some of the proposed strategies and ways of enhancing water efficiency include;

- Enhancing various flood irrigation mechanisms through reusing water runoffs, surge flooding, and improving canal linings
- Proliferation of various genetically modified plants that are resistant to droughts
- Shifting overreliance from "water hungry" fuels such as coal to other efficient energy source like natural gases
- Setting viable recommendation policies on biofuels
- Implementation of specific biofuels

POLITICAL SOLUTIONS

There are several mechanisms that can be adopted in order to help in securing safe and clean water for the people within nations. For instance, setting environmental policies and regulations that govern water usage within the boundaries of a country is one key strategy that can be adopted. Different strategies have been proposed by different governments in order to regulate and manage water consumptions and usage both within and outside nations' borders. The following integrated plan can be used for both developing and already developed countries.

Water Management System

To solve water security issues in the entire world, nations should ensure that people have access to sufficient safe and clean water at all times. There should be proper water management systems and proper supply and distribution methods in order to enhance water access for all people. Four key water management options have been proposed. These include;

- Water privatization
- Controlling all supplies and distribution systems (both locally and internationally)
- Pooling water resources through cooperatives
- Integrating all the three preceding options into one plan

Privatization of water: This is whereby all water management and distribution activities are given to private companies. In other words, private companies should be given full ownership of running and operating water business (including construction and maintain all the infrastructures used in water distribution systems). This helps in maximizing efficiency, especially in areas with severe water stress. Economically, private companies are given the mandate of researching and implementation of novel technologies in water systems in order to reduce the cost and improve the efficiencies in water distribution, thus more profits.

Although this is one of the key strategies that governments can adopt in dealing with the problem of water insecurity in a place, there are some shortcomings that are commonly associated with privatization of water distribution system. These include regulatory issues where private companies are fond of lobbing and putting strict standards to legislators. Additionally, private companies only focus more on urban and big cities because these are the areas with highest population thus translating to more profits. It has been revealed that out of the 1.2 billion people that lack access to safe and clean water and 2.5 billion people that lack safe sanitation worldwide, about 85% of them live in rural areas where water distribution is poorly done and managed. In other words, most private water companies usually do not have monetary incentives to cater and serve people from rural areas. Another issue with privatization of water management is that private companies are usually worried more about their shareholders rather than their clients. They don't care about basic human right (such as access to clean and safe water).

Governmental water control: This is another option solution that can be focussed on in dealing with water insecurity. In this case, the government is responsible for all mandates of ensuring that all people (both in urban and rural) have access to clean and safe water all the times at affordable price. One problem that faces this option of water distribution system is that it usually associated with a lot of inefficiencies. This is because the costs that the government charges to supply and distribute the water most often than not doesn't reflect the real cost of service. This makes water distribution system unsustainable. Lack of enough capital causes the government to fail to manage, improve, and maintain water distribution infrastructures. In case there are new technologies needed to enhance water distribution in a country, small government usually lack the needed capital to initiate and execute the project. Additionally, there is reluctance in public utilities in implementing new technologies or improving the quality of services in water distribution due to the bureaucracy that is involved.

Unlike the private water companies, the national governments should ensure that the right for people in accessing clean and safe water is adhered to. Governments should ensure that water distribution to all people is not violated even if it is not forthcoming in terms of returns.

Cooperative water control: This is another option of water securing. In this case, community members are usually involved in pooling different resources to develop water infrastructures that ensure access to clean and safe water for all. Cooperative water distribution and controlling methods are mainly practised in rural areas. This system is usually vulnerable and is open to all people. The key incentive to take part in this method of water distribution is access to fresh, clean, and safe water at low cost. Some of the main benefits of cooperative water control plan include;

- It helps in economy growth with a region through localizing all the money and resources
- Cooperative water control also helps in localizing political power and control
- It also encourage all people to take part in managing their water distribution utilities
- Generally, cooperative helps in offering water access that is reliable for all

There are also several challenges that are commonly associated with cooperative water control. Some of these challenges include obtaining adequate capital from the community members. It is also a challenge to educate all members in order to enlighten them on how cooperatives execute duties and operate. All member involved in cooperative water control should be enlighten or educated on how informed decision are made regarding management of cooperatives.

Trans-Boundary Agreements

One of key resource that is shared internationally is water. Internationally shared water resources are significant in promoting economic growth, human survival, production of food, and transportation among other benefits. Therefore, there is the need to always solve any crisis that might occur pertaining water security through technical solution. Different nations share water resources in terms of rivers, groundwater, and lakes. Conflicts between different nations may arise when they try to allocate water from these resources. Statistically, more than 250 rivers worldwide flow from one nation to another and about 45% of global population rely on freshwater from these shared rivers. About two third of these people are from developing and least developed countries.

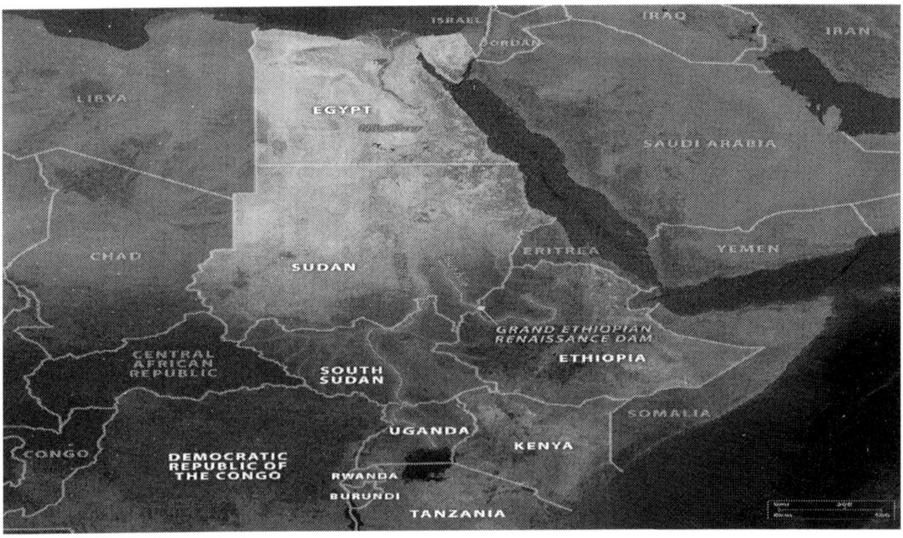

Figure 9.8: River Nile shared between three countries (Uganda, Sudan, and Egypt).

Source:*https://prezi.com/5grh1azymtsu/nile-river-basin-ggy-480-final-project/*

According to a study done in 2003 revealed that more than 1,800 internationally water related cooperation and conflicts occurred from 1950-2000. This shows that internationally shared water resources can act as a source of conflicts or can bring different nations together and promote peace through water treaty allocation which benefits all those countries involved.

Declarations

The right to water access is considered as one of key right to life for people. These rights usually offer an opportunity to practice freedom, fair justice, and optimal peace in the entire world. Access to clean and safe water is one principle that should be guaranteed in order to help in sustaining life and promoting prosperity and advancement in a nation. Therefore, different governments have a mandate and a duty to ensure these rights are promoted and protected amicably.

There are several declarations that have been set in order to promote and protect human rights to life and access to clean and safe water.

These declarations include *Universal Declaration of Human Rights of 1948, Stockholm Declaration on the Human Environment of 1972, Right to Development (41/128, 54/175), Bonn Declaration on Global Water Security of 2013, Human Right to Water and Sanitation (UN Resolution 64/292), UN Charter of 1945* and *Berlin Conference of 2004.* There are several declarations made based on clean and safe water access that have been formed as a way to help in solving crisis of water security in the entire world. These declarations include;

"The right to water is not an absolute right"

This declaration states that all water resources should be considered as part of the commons. In other words, water resource should not be owned by none, they below to all people. Water resources are renewable and right to access and use these resources should be absolute for all. However, this right does not means that people should misuse, alienate, or even waste these water resources.

"The human right to water"

This is one of the key and fundamental human rights. This right prioritise allocation of water for human consumption besides other uses. In other words, human beings should be given the first priority to water access. In this human right to water, there are several other small declarations that have been outlined such as:

- "The right to clean and safe water"-which states that human being should have access to quality water that does not affect their health status
- "The right to a fair minimum amount of water"- which states that human beings should consume on average 3-6 litres of water of a daily basis in order to sustain their lives. According to WHO recommendation, a mature human being should consume about 50 litres of water on a daily basis (these include water for cooking, food preparation, for sanitation, and for hygiene)
- "The right to accessible water"- this right states that people should be able to access water freely without travelling long distance. The minimum travel distance to water resource

should be 500 metres. The cost incurred to obtain the water should also be less than 3% the total annual income.

Right and Responsibilities of Governments

There are several declarations outlined based on this right. These declarations are discussed in detail as follows;

- "The right to development"- according to *UN Resolution 41/128* every human being has got the right to any kind of development. This declaration observes both country's rights to advancement and human right to development. This right must be uphold and respected by countries
- "The right to national sovereignty"- this declaration states that countries should not be challenged in terms of self-determination (including on issues pertaining water access). Additionally, this right gives a country the power to use any form of water resource within its boundary if the water distribution and supply systems fail completely.
- "The right to fair access"-according to Article ten of the Berlin Rules on Water Resources, a country has the right to access any water resources if the systems for water supply and distribution fail completely within the boundaries.
- "The responsibilities of a multinational world"-considering the right to fair access, this declaration states that a country have a responsibility to first consult with other countries before accessing shared water resources.
- "The responsibility to the people"- this declaration states that all people that are likely to be affected by a water project should be involved and made to participate exclusively in the process. Therefore, the government must enable those involved to have access to relevant information pertaining the water project

Protecting the Future

Proper implementation and management of this right should be executed in order to secure the right for future. Under this right, there are two key declarations that have been outlines. These include;

- "The responsibility to protect the environment"- this

declaration states that environments must be protected and preserved while executing this rights. This is because human beings dependent on the surroundings and therefore must be preserved and protected.

- "The responsibility to future generations"- this declaration states that it is the responsibility of governments and all people within a country to protect and preserve these rights in order to be accessible by other generations in the future.

Regulating Water Access through Environmental Priorities

Most countries in the world are still safe from different water crises that the governments have failed to handle. There are different governmental regulations that can be developed in order to deal with various challenges that contribute to water scarcity such as poor supply and distribution systems, diversions, and water pollutions. In various developed countries such as European Union and United States where environmental regulations have been effective for more than four decades, different legal decisions have always been made based on economic rationale.

One of the key strategy of establishing the most ideal and efficient standard to govern water security in a country is cost and benefit analysis. However, some places this strategy is not applicable because it fails to account for those that reap the benefits and those that bear the costs (for instance, when the U.S Environmental Protection Agency allows Tyson Fresh Meats company to disposed all its animal waste (nitrites and nitrates) in to the Missouri river according to cost and benefit analysis, there limits that are set at the "economically efficient" amount) in this case the total peripheral costs incurred to lower total amount of animal waste that is disposed into the river traverses the peripheral social cost incurred to disposed the waste into the river.

Regrettably, all those people that use the river water downstream are the ones that deal with all sorts of challenges related to contaminated water (including unpleasant smell, dangers of nitrites and nitrates, sights, and other menace of the river water) and not the company dumping its waste effluents in to the river. The company (Tyson) does not receive incentives from the government in order to reduce the danger associated with water pollution. This is because this is not one of its economic priorities. In other words, the company does not get or reap any benefits from reduction of pollution in the river. This aspect where one party is directly affected by economic decisions that are made by another party is commonly referred to as externality.

Realistically, most of those powerful and developed companies that have achieved their economic success have the ability (financially) to create novel ways and technologies that can help deal with water security of a place, but they are usually the last in terms of making financial decisions towards innovative technologies needed to solve a problem that they have little or no need for it. Such companies can be involved in innovative approach of dealing with water crises in a country through the use of environmental regulations.

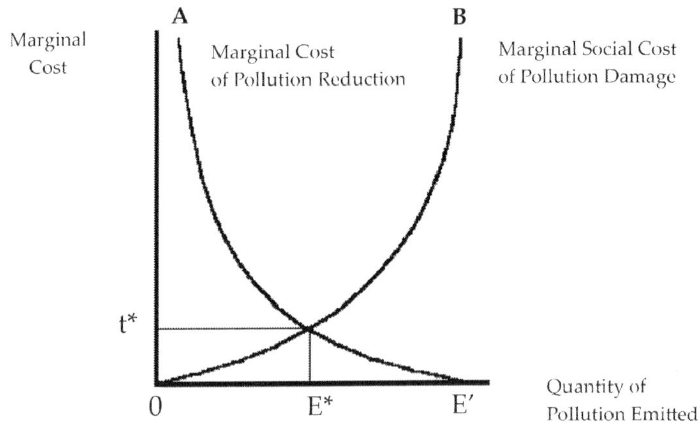

Figure 9.9: An environmental regulation calve: E* represent the point referred to as "economically efficient".

Source:*http://12.000.scripts.mit.edu/mission2017/regulating-with-environmental-priority/*

The government should ensure that there are environmental regulations set to control diversions, pollution, and distribution of water in the country. This enables all people to access sufficient clean, safe, and fresh water for both present and future use. In case these environmental regulations exist, the governments have a mandate to disassociate them from various economic regulation rationale methods, and align them in accordance with principles and goals set of supplying sufficient water to societies, while at the same time making sure that all the natural resources exist and are safe down the road. This should also ensure that the next generations have access to reliable and safe water resources in future.

Water Security in Developing Countries

Most developing countries have numerous challenges when it comes to handling water security issues. They lack the means to solve and fix various water related crises. There are three key categories of water related challenges associated with developing countries. These categories are outlined below as;

Access: most people in developing countries such as in Africa and some parts of Asian countries have limited access to safe, clean, and sufficient water. Most of these countries face different problems in solving issues of water crises. This is because of lack of adequate funds and means to solve the problems.

Management: there are also few governmental regulations that are set aside to deal with water security in developing countries. To help in the implementation of governmental water regulation in developing countries, there is the need to adopt integrated water management plans.

Agricultural productivity: most of the water in developing countries is inefficiently used in various agricultural activities. One of the key strategies that governments from various developing countries can adopt is the use of genetically modified crops which are efficient in terms of the total water used to grow them till they are ready for harvesting.

It is the responsibilities of governments to try and regulate various companies involved in biotechnological activities in order to prevent and curb most of the potential challenges faced through big corporation invasion in developing countries. To do this successfully, there are three key recommendations that every developing country should follow. These recommendations should be used together for them to be effective.

- The use of multilateral economic groups: developing countries should try to capitalize on various groups such as central banks and world banks in order to obtain funds for financing various water related projects.

- Micro-financing: this is one of the key strategies that developing countries can use in loaning citizens small amount of money. This method is useful in alleviating various symptoms of water security in a country.
- Charity organisation: this can be used as short-term solutions for water related crises in developing countries. A country can also benefit from international governments in terms of charity funding, grants, and relieves to solve various water related challenges.

ECONOMIC SOLUTIONS

There are different significant methods that governments can undertake to help in enhancing water security economically. One of these methods is making sure that all involved persons, corporations, businesses and governmental agencies involved in water distributions are given economic incentives that will enable them meet the required standards in provision of clean and safe water for all. There are several plans that can be developed to help in creating economic incentives in order to meet these standards. These plans should effectively solve water security issues in both developing countries and those already developed countries suffering from different violations on rights to access safe and clean water such as United States, China, and India.

Water Tariffs Plans

Currently, many developing and developed countries are facing different challenges in clean water pricing and distribution systems. Some of these problems include wastage in water use, scarcity of safe and freshwater, and lack of sufficient and affordable water.

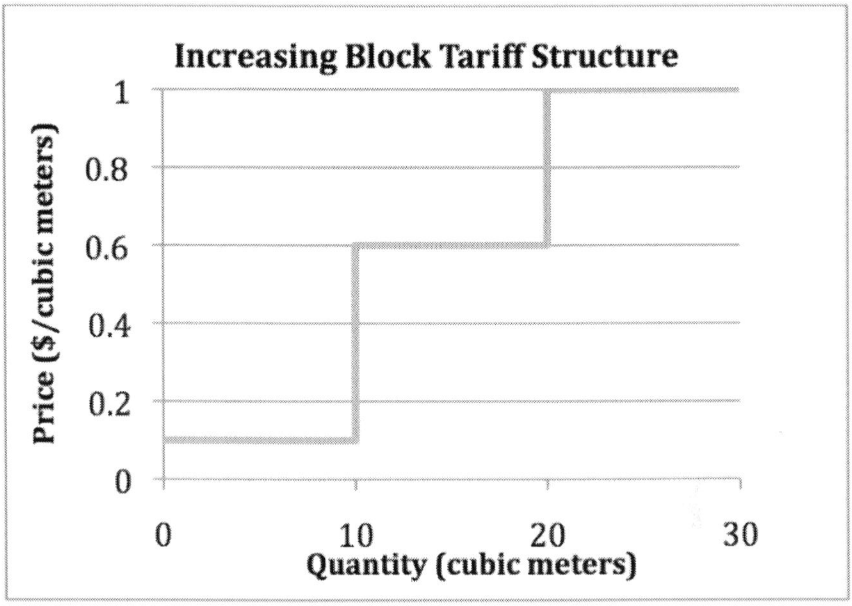

Figure 9.10: Increasing block tariffs.

Source:*http://www.sswm.info/content/water-pricing-increasing-block-tariffs*

One of the key solution to these problems is aggravating block tariffs. In this case, water usage is divided in terms of blocks or tiers, where pricing per water unit is aggravated based on the average consumption rate of the water. In the first amount block, the price of water should be lower. In the second block of water usage, pricing should be done based on the rate of consumption in that block. This should go on until the last block. There are three key features that should be adhered to in each aggravating block tariff. These include;

- The total number of blocks that exist in a water distribution system,
- The total amount of fresh water that is used in each block,
- The price for each water distribution block

These three features are usually influenced by the amount of water that is consumed in definite region. The initial water usage block (which is also referred to as lifeline or social block) should be inexpensive in order to cater for the needs of all households. In other words, the price in the first block should be set below the actual water price. Other blocks should be set in order to act as incentives to compensate and conserve the price for the first block. Most poor people especially those in lower income bracket usually benefit a lot from the subsidized cost of water in the first block. Those that use more water are required to pay a higher price in the higher water block.

Proper implementation plan for increasing block tariffs is important in order to enhance water distribution system. The total number of water blocks should be governed by the consumption rate and pattern of water of industries or regions. All prices set for the domestic water usage should be lower compared to the industrial water usage. The cash that is obtained through water supply to different customer should be used for water conservation purposes. Alternatively, this money can also be used in subsidizing water distribution activities for the poorest domestic households.

When implementing increasing block tariffs, it is advisable to always price the water based on the total number of people per household. Total population per household in developed countries can be determined using tax report information. In least developed countries, obtaining tax report information may be difficult. Therefore, increasing block tariffs would be applicable for the developed and developing countries, especially those that have advanced infrastructures that can be used in obtaining relevant information for every household based on the total number of people and those with special needs. Those nations where implementation of such systems is applicable, every household should be able to pay their water usage based on the increasing block tariffs and the characteristics of their household. For instance, a household with six members should be in a position to pay double the price payable by a household of three members in the first block. Regrettably, those countries without proper infrastructures will be required to deal with water security problems until they are able to acquire the needed information systems.

Cap and Trade Plans

Cap and trade plans are environmental policies that are set in order to help in reducing various pollutants from industries such as sulphur dioxide and carbon monoxide. Governments are responsible for setting key caps that should govern the quantity of emissions into the environment. Industries are given allowances or permits based on the caps set. These allowances are usually auctioned, free, or even both. To deal with water pollution, governments should determine amount or the degree of pollution and then issue permits based on the level of threat they pose to water for domestic use. The responsible agencies and authorities can use the list of conventional pollutants outlined under the "Clean Air Act in the U.S. under the Appendix A to part 403, reprinted after 40 CFR 423.17" as a guideline to determine the level of threat posed by their own conventional water pollutants. Some of the benefits of using cap and trade strategy to solve water security challenges in a country include creation of new markets in a country, reducing the level of pollutions in water resources, and help in revenue creation for the authorities involved.

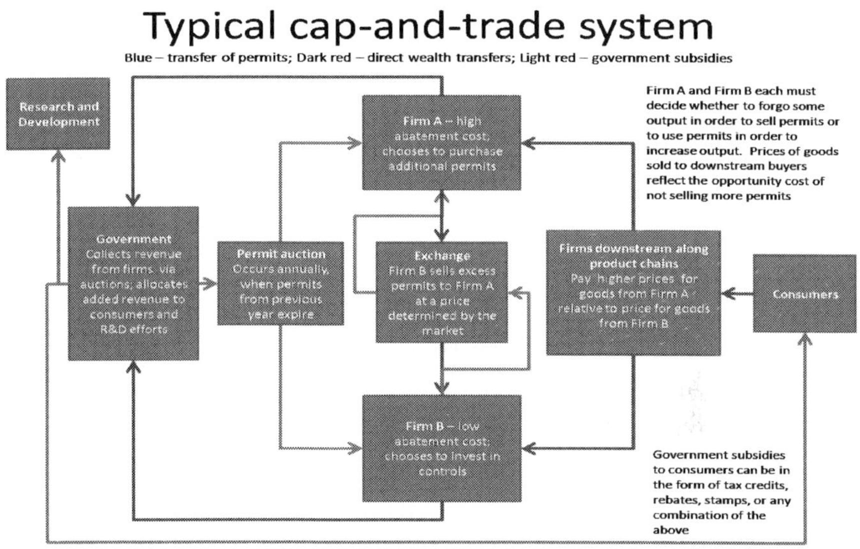

Figure 9.11: A typical cap-and-trade system.

Source:*http://web.mit.edu/12.000/www/m2012/finalwebsite/images/cap-and-trade.jpg*

There are also some shortcoming associated with the use of cap and trade in solving water security related issues. For instance, it is not applicable for the countries without central governments. Cap and trade may also affect domestic companies negatively in the international trades. This is because the additional capital that domestic company incur in implementation of cap and trade. Lastly, cap and trade set to deal with pollution on freshwaters may also inadvertently incentivize different companies to pollute water through discharging their effluents in to freshwaters.

SOCIAL SOLUTIONS

The issue of water insecurity in countries should be dealt by both the government and all the people in those countries in a holistic approach. There are different plans that can be developed in order to solve challenges of water security both in developed and developing countries. These plans are well outlined below to help the reader obtain the necessary information required to practically execute those plans.

Virtual Water Labelling Plan

Countries should ensure that conservation of water resources is attained in order to protect and preserve freshwater water supply globally. This can only be achieved if all the people cooperate with various agencies and authorities involved in water management and preservation. To enhance this cooperation, the government should try all it can to create awareness among citizens on the benefits of conserving water in everyday activities.

Using virtual water labelling strategy, agriculture is one of the key areas that should be targeted first. This is because studies have shown that agricultural activities account for more than 75% freshwater usage in the world. There are several ways that farmers can use to reduce agricultural water usage. These include using genetically modified crops, using proper irrigation methods (such as drip irrigation), fertilizer monitoring, and the use of pesticides. Agricultural companies should also be encouraged to participate in virtual water labelling plan.

Governmental legislation should also be enacted that require all farmers around the world to participate in this plan by calculating and labelling the virtual water in their farm products. All farm produces usually have a level of monetary value. Therefore, this value should go along with the virtual water volume based on this plan.

Various companies that are involved in categorizing farm products using their virtual water use product water footprint based on three key colours-that is blue green and grey. According to the figure below, the blue water colour represents water from all fresh water sources such as aquifers, lakes, and rivers. The green water represents all water from precipitation which does not form runoff. Finally, the grey water represents the freshwater that has become polluted in one way or another. Based on these three types of virtual water, it is important to always distinguish the type of virtual water on labelled on different products.

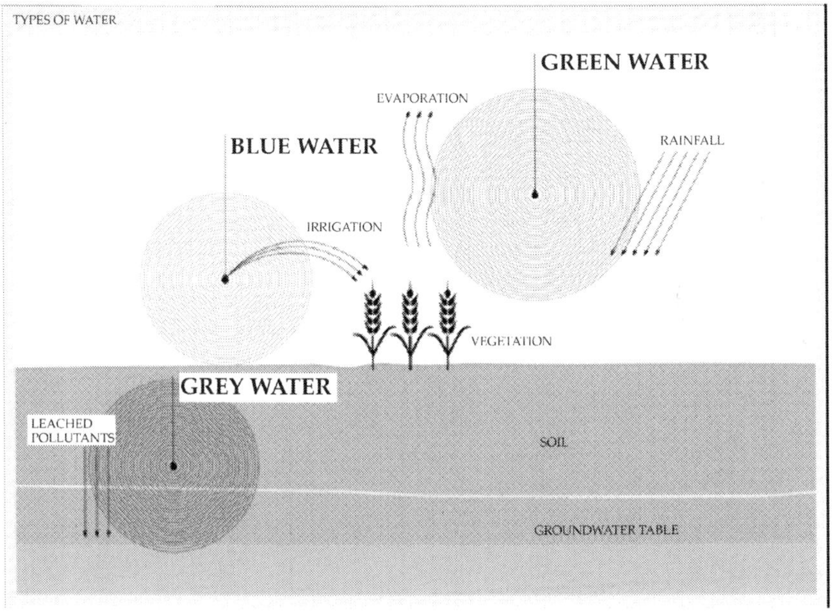

Figure 9.12: Virtual water labelling plan (Green, blue and grey water presented as part of the water cycle).

Source:*http://12.000.scripts.mit.edu/mission2017/solutions/people/virtual-water-labeling/*

Of the main challenge of using virtual water labelling plan in water security is that it sometime becomes very difficult to monitor various methods that are used by farmers and their accuracy in obtaining the best results. This method also does not guarantee that customers will shift their demand on certain products based on the use of virtual water labelling. To help deal with these challenges, government should step up and develop ideal methods to calculate the virtual water required.

Social Solutions to Energy and Water Problems

Energy production is another key sector where water usage is high around the world. Studies have revealed that more than 4% of all freshwaters in the world are used in energy production processes. This means that the more people use energy around the world, the more freshwater is used in production of that energy. Production of certain energy such as ethanol commonly use more water compared to other conventional types of energy productions. It is therefore important for various policy makers from different countries to come up with most appropriate decisions on how to effectively allocate water more efficiently for energy production especially in those regions that experience prolonged water scarcities. There should be a balance between water utilization in energy production and other domestic and industrial uses. This will help in minimizing the conflicts between water and energy throughout the world because human beings require both water and energy for their survival.

Beside conserving and protecting water resources, it is important to also try educating people on how to reduce consumption rate of energy in general. This will in turn helps in reducing water use around the world and lower the conflict that may arise in future between water and energy. With the current trends in global energy use, it is projected that this use may go up by about 55% by the year 2040. Additionally, the figure below shows that demand for water in the world is likely to go up, thus resulting to huge problems and conflicts between water and energy in future.

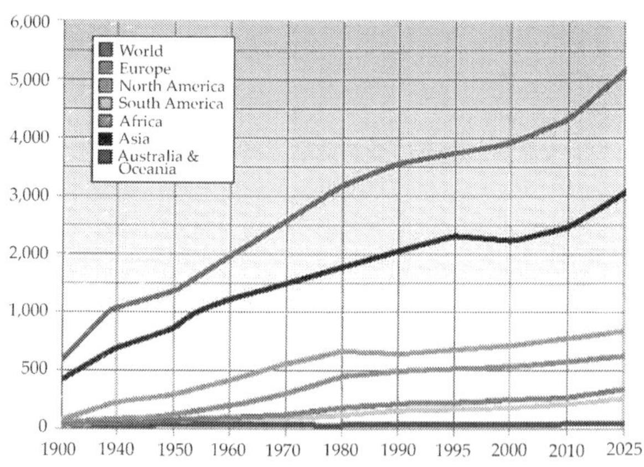

Figure 9.13: Global water consumption 1900-2025 (by region, in billions M^3 per year).

***Source:**http://12.000.scripts.mit.edu/mission2017/social-solutions-to-energy-and-water-problems/*

Governments are encouraged to seek other energy production alternatives in order to conserve and protect water resources both locally and internationally. There is the need to also look for other means of transportation that consume less energy. This in turn helps in conserving the water use in energy production. One of the key energy production alternatives locally is solar energy. It has been proven that the use of solar power is likely to reduce consumption of water by about 70-80%. The use of solar energy will not only reduce water consumption, but it is also safe and clean to the environment.

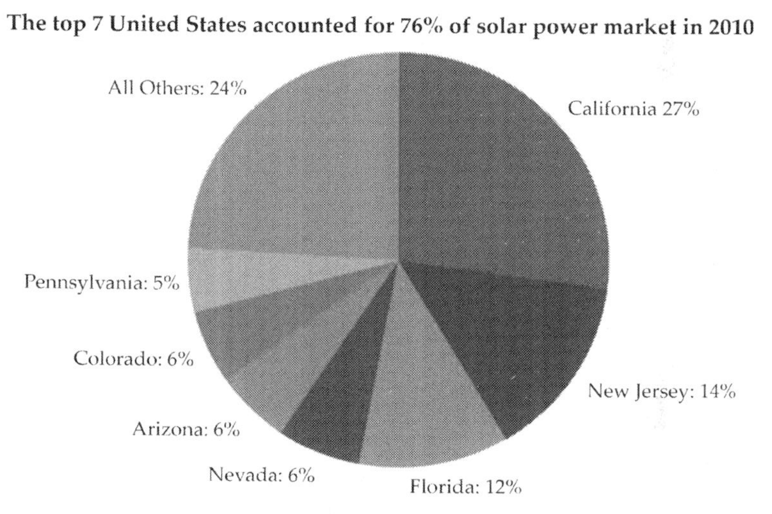

Figure 9.14: Solar Power installed in United States by 2010.

***Source:**http://12.000.scripts.mit.edu/mission2017/social-solutions-to-energy-and-water-problems/*

People will also tend to lower their energy consumption rat if they notice that they are using more energy uneconomically. This scenario would cause many people to shift from using costly energy to other renewable forms of energy that they believe to have fewer impacts on the environment and water usage. For instance, in the figure below, United State government subsidize the price of E85 ethanol. This made the fuel to be more competitive with other forms of fuel such as gasoline. Despite the fact that water usage for ethanol product is higher compared to other conventional biofuels, the subsidy is used to make ethanol to be more competitive in the world market with other fuel such as gasoline.

Statistically, ethanol is considered one of the key energy in United States. It actually represents about 10% of the total market share of the country's gasoline. In 2010, the total amount sold was 13.2 billion gallons compared to the 1.63 billion gallons that were sold in 2000. From this example, it can be deduced that the action taken by country's government can seriously impact on the citizens' behaviours. Therefore, it is the responsibility of every government to try and make fuels that use less water and are environmentally friendly to be economically competitive.

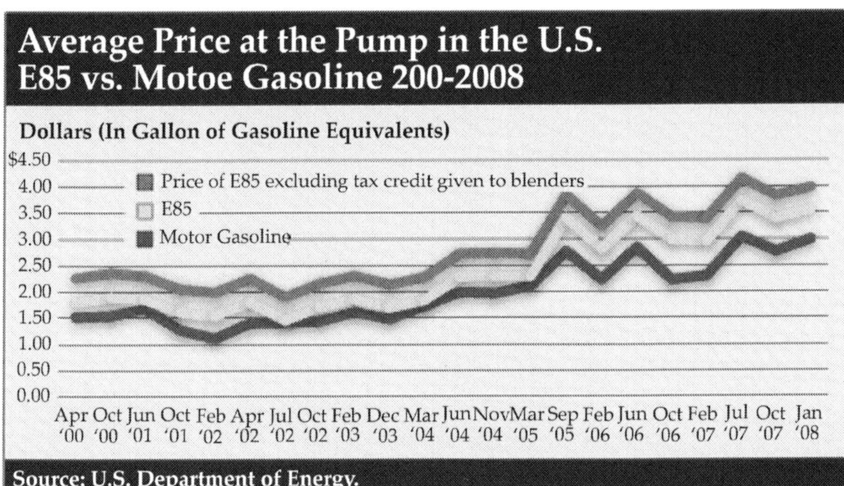

Figure 9.15: The price of ethanol at the pump in the United States verses motor gasoline 2000-2008.

Source:*http://12.000.scripts.mit.edu/mission2017/social-solutions-to-energy-and-water-problems/*

In a summary, the government can come up with specific programs to educate its citizens on the importance of energy conservation. These programs should be tailored towards reduction on energy consumption rate and water usage. World leaders should also be enlightened on how to make wise decisions for their people. They try to use energy production processes that consume less water especially in countries with scarce water sources. International organisations also have a responsibility of creating systems that will help in advising different countries on the most viable types of energy to use.

Community Involvement and Education

According to the declarations outlined in the previous subtitle, it was mentioned that every country should try by all means to promote and protect peoples' rights to access clean and safe water. Recently, water access for all has become the talk of the day thus labelled as a UN Millennium Development Goal. Educating and involving community to participate in water conservation and protection is a key strategy towards dealing with water security issues not only at a national level, but also internationally.

When dealing with water security issues in the country, there are several principles commonly referred to as "Dublin Principles" that can be used. These principles help in outlining and solving various issues of water security. They provide a platform for potential solution to water crises. They are outlined as follows;

- "Water is a finite, vulnerable and essential resource which should be managed in an integrated manner"
- "Water resources development and management should be based on a participatory approach"
- "Women play a central role in the provision, management, and safeguarding of water"
- "Water has an economic value and should be recognized as an economic good, taking into account affordability and equity criteria"

These principles usually act as a guideline on how water resource should be handled. They also act as platform for Integrated Water Resource Management (IWRM). There are three key water management goals that are usually addressed based on IWRM. These include;

- Ecological sustainability,
- Economic efficiency,
- Social equality

For instance, based on these three water management goals, different countries such as Singapore have successfully management water security issues through implementation of different solutions that are similar to those of Integrated Water Resource Management. Several programs were put into practices in Singapore such as increasing community participation in water management, assigning water resources an economic value, recognizing the influence of households in water management activities, and respecting and protecting water as one of the key finite resource. For a long period of time, these programs have help Singapore and other countries to remain water secure nations.

Despite the benefits that come with application of Dublin principles in water management, most developing countries still suffer from poor community participation and empowerment due to lack of proper education. Those people at community level that suffer the most (women, children and the poor) usually do not have any political influence. They suffer from poor water governance. To improve on their participatory role and influence, they should be involved politically within their communities.

Education will help people be able to see abuses, argue for water reforms, and speak out their minds. There should be open communication between the local and national governments in allowing citizens to take part in management of local water resources in order to solve various issues experienced both at community and national level. In other words, people who are given a chance to exercise their power and knowledge in order to enhance their lives in a better way should be considered as a key step towards creating a society that is empowered and committed in dealing with water related crises. Enlightening people on how to take part in promoting and conserving water resources can help in preventing water abuse and contamination. This allows people to have an access to safe, clean, and reliable water supplies.

Education can also help in enlightening communities about different water efficiency and health issues. Studies have shown that more than 40% of people from least developed countries have little or no knowledge regarding various water-borne infections such as typhoid and cholera. Education can therefore help in enlightening people on how to invest wisely on water treatment projects. It can also help people in differentiating between potable water and non-potable water sources. Water demand can be reduced through recycling waste waters, proper waste disposal, crops maintenances, cleaning, and obtaining safe water for domestic use. More than 70% of water that has been used for a specific role such as hand washing can be reused in crops watering or toilet washing.

Women should be involved in water management practices because they act as key pillars of water supplies in most households especially in region with scarcity. Women and young girls in developing countries spend most of their times searching for water. Involving young girls at school in education pertaining water management is vital towards enhancing water security in a country. because most of activities involved in water collection, cleaning, waste water disposal and cooking are done by women, it is crucial to offer them the best education that will help them manage and protect water resources to have secure supplies for current and future generations. This education should be tailored towards ensuring that women are able to reuse recyclable waste water, proper dispose waste water, and management water properly.

To help in implementation water management practices, the government should try to include some topics in school curriculum that will enlighten the students on water sustainability practices. Students should be taught about various consequences that may arise if water resources are abused, and how they can take part in ensuring that these water resources are protected and conserved properly. Students that take part in enhancing water security in a country should be given incentives in order to boost their morale and that of others in management of water resources. Such an initiative has been practised successfully in some countries such as Singapore, where education at community level has played a significant role in reducing water demand per person per day by almost 20 litres.

When developing infrastructures for water supply it is important to ensure that community members are also well trained on how to protect and maintain all water resources used. Studies have shown that piped water may be waste through leaks which may account for more than 60% of water wastage. Therefore, members of the community should be trained well on how to handle such issues whenever they arise. This does not only help in managing and protecting water waste, but also helps in job creation in communities. The government should also come up with special programs at community level that helps in educating key persons who will take part in enlightening other members of the community on how to properly manage and protect their water sources. Apart from country's government, other humanitarian groups also play a key role in funding and promoting education in communities on how to secure water resources.

CHAPTER 10

DESALINATION FOR SUSTAINABLE WATER SUPPLY

INTRODUCTION

Desalination is described as the process of mineral extraction from the sea water. Generally, sea water is saline and unfit for human consumption. Therefore, the salt and other minerals are usually removed thus making the water fit for both human consumption and for irrigation. The idea of desalinating seawater is to improve safe and clean water supply, especially in those countries such as Middle East faced with a lot of challenges in water security.

DESALINATION TECHNIQUES

There are different techniques used in desalination of seawater. The choice of technique would depend on a number of factors including salinity or brackishness of the water, scale of water requirement, level of availability of skills to operate and maintain the plant, energy use and power sources available, capital costs, and operating costs among other factors.

Brackishness of Water

Salinity is determined by the amount of dissolved salts in the water. It is usually indicated based on the Total Dissolved Salts (TDS) value. According to World Health Organisation recommendation, Total Dissolved Salts values of between 100 and 1500 milligrams per litre for drinking water is considered the best for desalination of seawater. However, in those places that water supplies is not a problem, a narrower range of 300 to 800 is more typical preferred. Brackish water has Total Dissolved Salts values in the range 1000 to 5000. For moderately saline water and severely saline water Total Dissolved Salts values are 5000 - 10,000 and 10,000 to 30,000, respectively. Typically seawater has a Total Dissolved Salts value of about 35,000, which is considerably higher than most inland sources, though not always the case. There are several guidelines on the typical ranges of Total Dissolved Salts values for water which can be processed with different techniques. These guidelines include;

- Ion Exchange - 500 to 1000
- Electrodialysis - 500 to 3000
- Reverse Osmosis (standard membranes) - 500 to 5000
- Reverse Osmosis (high resistance membranes) - over 5000
- Distillation – 1000 to 100,000 +, but especially over 30,000

Additionally, electrodialysis and reverse osmosis require water that is otherwise clean, apart from containing dissolved salts. These techniques would do little to remove bacteria from water, and particulate solids would be especially damaging with reverse osmosis as they would clog up the membranes and make them ineffective. To remove solids and bacteria from water, additional processes would be required before the electrodialysis or reverse osmosis stage. In contrast, distillation would separate out water from solids, but the solids would need periodic cleaning out, and it can help to remove bacteria provided care is taken to keep the water input separate from the distillate.

A reservation on electrodialysis needs to be mentioned in the case of relatively poor and isolated communities. For such communities, although electrodialysis is cheaper and less complicated than reverse osmosis it might not be a realistic option especially if the Total Dissolved Salts value is above 3000, distillation would be much more effective. If it is below 3000 and in the range for which electrodialysis is normally used, local people are likely to be used to brackish water and might not consider it a priority in development terms to get treated water. In reality most people become accustomed to their source of drinking water, even if rather brackish and possibly causing some longer-term health risk. It is likely that in some areas people are making use of water with Total Dissolved Salts values even in excess of 3000.

Water Scale Requirement

Personal consumption of water is very variable. An absolute minimum in hot climates, for survival is about two liters per day as drinking water. This leaves no water for personal washing, washing of clothes, cooking and cleaning of eating and cooking utensils, etc. A more realistic minimum water requirement per person is therefore generally taken as about eight liters per day. However, many millions of the poorest people, especially in arid or semi-arid areas have to manage on less than this amount of water for significant parts of the year. A figure for personal consumption generally taken to indicate absence of a shortage of water is 20 liters per day. Of course in countries in the West many people are extravagant in personal consumption of water and figures in excess of 200 liters per day are not uncommon.

Therefore, a household of, say, eight people would have a water requirement of 160 liters per day, say 200 liters (0.2 cubic meters) for a bit extra. For a small settlement of about 100 people the requirement would be about 2000 liters (2 cubic meters), or a little bit over, and for a larger settlement of 500 people the requirement would be 10,000 liters (10 cubic meters) per day. If the desalination unit is based around a water pump or well, a typical output would be up to 5,000 liters per day with manual water lifting, although the output from a powered pump would be considerably higher. A manual pump or well would therefore be comfortably used by no more than 200 or 300 people. A desalination unit on about this scale would then also be appropriate. Of course a standard water pump can be used by a community of less than 200 to 300 people, in which case its output would be lower. However, if there are too few people using the facility, probably less than 100 in many places, it would be too difficult for a small number of relatively poor people to be able to maintain a pump.

Below are given approximate maximum and minimum production rates with various desalination technologies. The consumption based on the number of people using the facility and personal consumption of 20 liters per day can be used to size the facility, then a technology chosen for which the total consumption fits within the specified range.

Table 10.1: Illustrating approximate maximum and minimum production rates with various desalination technologies

TECHNOLOGY	MINIMUM PRODUCTION – M³ PER DAY	MAXIMUM PRODUCTION – M³ PER DAY
Solar stills	0.002	200
Electrodialysis	0.1	15,000
Reverse Osmosis	0.5	100,000
Multi Stage Flash Evaporation	4,000	45,000
Multi Effect Distillation	1	10,000

Source:*https://knowledgepoint.org/upfiles/14466282963061441.doc.*

It needs to be noted, though, that reverse osmosis, multi-stage flash and multi-effect distillation are relatively complex technologies the cost per unit output of which would increase sharply at the smaller scale of their range of operation, so smaller scale operation might only be viable in particular situations, e.g. where there are no other viable sources of water and a settlement is small but relatively wealthy, for example on the Gulf of Arabia.

Operating Skills

All types of desalination technology would require people with particular degrees of skill to manage and maintain them. Medium or large-scale reverse osmosis and electrodialysis plants and plants based on multi stage flash and multi effect distillate would require a dedicated team of trained technicians, mechanics, electricians, engineers and managers in the same way as any other type of technology-intensive industrial process would. It could, however, be of interest to consider the possibilities for community-based maintenance, possibly linked to some form of more specialist back-up – e.g. by a local research institution or private company, in the case of small-scale electrodialysis, reverse osmosis or solar distillation.

Operators of electrodialysis and reverse osmosis plants would need to have a range of technical skills e.g. understanding of electrical circuits, repair of pumps, or at least knowing what the problem is when pumps don't work, ensuring proper operation of filters, dismantling of membrane units or electrodes for cleaning or replacement, maintenance of battery units, maintenance of photovoltaic arrays, wind turbines or generator sets or, particularly in the case of reverse osmosis, hydraulics and repair of leakages, especially at joints in areas of high pressure operation. The operators would also need to have a realistic view of the cost of the services they provide and the costs of operation, management, maintenance and repair, and a process needs to be in place of recovering these costs from the users, who themselves would need to be in a position to pay these costs, which could be impossible if many of the users are very poor. Here a specialist local institution could be of assistance. It could undertake a preliminary draft cost analysis at the planning stage to decide whether a project is viable and identify or assess possible local expertise to run a plant.

There is some experience from India of small to medium-scale (10 to 300 m3 per day) operation of reverse osmosis plants in rural areas. Desalination plants should be set up in partnership between national and state governments, research institutions, private companies, which provide most of the plant, and local organizations. Plants are also set up where there is usually a cluster of villages dependent on a brackish water supply, and water is piped from a central plant to households or standpipes in the villages. Local people should be trained in day to day running of the plants and routine maintenance. More extensive repairs require an engineer or technician from a private company or research center to be called out. This cause delays and non-operation of the plant, especially due to shortage of spare parts. It is not known if there are any operational reverse osmosis plants of smaller than $10m^3$ are operating anywhere, other than as pilot or test plants.

Solar still plants are technically much simpler than other types of desalination plant. Most significantly, the still units themselves contain no moving parts – though pumps might be involved for water lifting from the ground or to a storage tank or reservoir. It would be reasonable to assume that solar still plants could be operated and maintained largely by communities themselves, even if they contain no-one with some technical education or experience. Availability of skills would therefore not be likely to be critical factor in determining the viability of an operation, and factors such as levels of salinity and brackishness, suitability of other types of water supply and availability of construction and repair materials would be likely to be more critical. There is similarity between small-scale water pumping technologies and desalination using solar stills in terms of sustainability of operation. Small-scale water pumping is a mature technology which has much wider application than solar desalination or distillation with regard to the suitability of particular pumping systems to operation and maintenance in rural communities.

Requirements and Source of Energy

Distillation is an energy intensive technology, as energy is required to evaporate water. Reverse osmosis uses much less energy, the main requirements being largely for applying pressure (between 15 and 100 atmospheric pressure depending on the salinity of the solution and type of membrane used, for overcoming the osmotic pressure of the solution across the membrane) and to pump water around the system. Electrodialysis generally needs more energy than reverse osmosis, despite not requiring to be operated at very high pressures. This is probably due to the inherent inefficiency of the process making it necessary to pass the solvent through the unit at least three times, whereas the efficiency of the reverse osmosis process is much better. Also, for electrodialysis there is a need to heat the input solution to 30 to 35°C in order to improve ion removal efficiency, which also adds to the energy requirement. The table below given a comparison of energy requirements based on the various processes of desalination of seas water. These figures are only quite crude as direct equivalence is not generally possible, with actual consumption depending on very many factors e.g. ambient temperature, wind speed, altitude, size of plant, compactness of plant, extent of heat or pressure recovery incorporated, age of plant and how well it has been maintained.

Table 10.2: Illustrating a comparison of energy requirements based on the various processes of desalination of seas water

Process	**Total Energy Consumed – kWh/m³**
Multi stage flash (distillation)	60 to 70
Multi effect distillate	About 50
Multi effect distillate with vapor compression	About 10
Solar still	About 1000
Reverse osmosis	4 to 10
Electrodialysis	About 10

Source:*https://knowledgepoint.org/upfiles/1446628296306144l.doc.*

One feature to note about the above table is the very large figure for energy consumption with solar stills. Distillation without any heat recovery as the distillate condenses is not an efficient process. Its attraction is the direct use of solar energy which is, in effect, free energy. For this reason distillation by direct application of heat from a fuel source on a small scale is not an attractive option. In some areas where water is brackish or saline indigenous stills fired with wood or charcoal are sometimes used, for instance in Botswana. Previous dissemination of higher yield fired stills based on traditional models in Botswana had resulted in rapid depletion of firewood reserves around the villages where they were used. The decision was then taken to develop solar-based stills. The larger-scale industrial distillation processes (multi stage flash and multi effect distillate) incorporate recovery of the heat as the distillate water is condensed. With vapor compression much of the energy used to evaporate the solution is recovered when the vapor is compressed making the process highly efficient.

Multi stage flash and multi effect distillate utilize electricity from a generator or grid system or steam from a separately fired industrial boiler. Most of the energy requirement is to raise steam which is needed to drive the process. At present it would be likely that only desalination plants of these types at the smaller end of their ranges could be powered wholly or in part by energy from renewable sources. Solar stills, on the other hand, by definition use renewable energy. They might need to incorporate pumps to extract water from the ground or reservoir or to take treated water to a tank or reservoir, but with small-scale systems the pumps can be manually powered or use electricity from solar photovoltaic or wind sources. Solar stills can have a fan installed in the unit to increase output and so reduce the area needed to be covered by solar collectors.

With reverse osmosis and electrodialysis the main energy requirement is electricity for pumping and also, in the case of electrodialysis, to maintain an electric potential gradient to cause ion migration out of solution. Also with electrodialysis, a separate heat source might be required in cooler climates to raise the temperature of the water input to above 30°C at which temperature the efficiency of the ion removal process improves. There is therefore considerable potential for use of renewable sources – solar photovoltaic and wind, to provide some or all of the electrical input. A number of organizations have been developing and trialing relatively small systems, generally of 0.5 to 3 m^3 capacity, powered by renewable energy. In 1995 an electrodialysis unit of approximately 3.5 m^3 capacity per day, powered by solar photovoltaic or wind at Spencer Valley, New Mexico, USA, constructed for Navajo inhabitants had been in operation for three years. A fairly large dispersed wind-powered reverse osmosis plant was built in 1998 on the island of Syros in Greece. This contains several reverse osmosis modules near the sea and a series of wind propellers and a wind turbine on a hilltop site 1.5 kilometers away. This produces between 60 and 900 cubic meters per day, depending on wind conditions. So far, this has been a one-off plant and the technology involved relatively complex, so it is not known if such a plant could become more widespread.

Installation and Operation Costs

If there is a problem of brackish and saline water in a particular area, desalination would not always be the most economic option. Other possibilities could be; bringing in water from another area by vehicle, supporting informal processes already operating (such as water carriers who make a living from supplying water, and extending their scope), drilling a deeper borehole if it is known that water at greater depths is cleaner, improved rainwater harvesting and storage, piping in water from more distant sources, and using a comprehensive water improvement program if the area affected is large and incorporating one or more of the above (which might also include building of water storage tanks or reservoirs).

The best option would be very dependent on local conditions and factors, and these would need to be assessed when attempting to make estimates of costs for various possible options. According to Commonwealth Science Council review points, it has been revealed that in industrialized countries a desalinated water supply is 2 to 50 times more expensive than water from other sources. However in some areas in developing countries, where water shortages can be acute, the cost difference might not be so great. Desalination at small or medium scale could offer particular advantages if;

- The community is relatively self-contained and not highly dependent on interacting with other communities
- Alternative water sources are impracticable (such as very distant, very deep underground or rainfall is very limited). This might be the case with some relatively dry islands where the only realistic option to supply water for most people is seawater desalination
- There is already some desalination being undertaken elsewhere in the vicinity
- There is political, and possibly donor support to enable testing and optimizing operation, possibly initially as a subsidized operation
- Local skills can be readily adapted to building, operating and maintaining desalination units
- More specialized technical support is available, e.g. from a university, research center, private company, NGO or government agency
- Materials for building and spare parts can largely be sourced locally.

If it is decided to go ahead with a desalination option then to a large extent the preceding considerations would determine which would be the most suitable type of process. By way of comparison between the processes the following is a very approximate estimate of costs.

Table 10.3: Illustrating a comparison of costs based on the various processes of desalination of seas water

PROCESS	INVESTMENT COST US$ PER M³	OPERATING COSTS US$ PER M³	TOTAL COSTS US$ PER M³
Multi stage flash – about 50,000 m³ per day	About 0.31	About 0.74	1.05
Reverse osmosis - about 50,000 m³ per day	About 0.30	About 0.64	0.95
Solar still			About 25
Multi effect distiller			About 10
Electrodialysis			0.25 to 1

Source: *https://knowledgepoint.org/upfiles/1446628296306144l.doc.*

IMPORTANCE OF DESALINATION

Desalination is becoming increasingly important for water supply. However, this has largely been through technological development of large-scale plants and the growing importance of desalination in particular situations. Among these situations include arid and semi-arid areas experiencing relatively rapid population growth, but where, nevertheless, the majority of the population cannot be considered as poor (for instance on the Arabian Gulf and some other Middle Eastern and North African countries), Tropical islands, where water from rainfall and other natural sources is not adequate to supply steadily growing populations and tourist influx (for instance on Canary Islands, Aruba, Corfu – Greece, the Virgin Islands, Antigua, the Cayman Islands and Malta), Reservations of native people, mostly in the USA and Australia, Areas of countries in the North (such as the USA, Spain and Greece) with seasonal or erratic rainfall, relatively wealthy stable or growing populations, growing personal consumption of water and expectation of water availability, and damaged or vulnerable eco-systems which can no longer support the same level of accumulation of water as they once did. Common factors that affect the extent of desalination use are increasing demand and expectation, lack of suitable alternatives and a willingness and capacity to pay comparatively high prices for water supplies.

In developing countries in many areas there are more compelling water needs due to people attempting to make a livelihood from increasingly marginalized land, increasing desertification and ecological degradation, increasing numbers of people having access to insufficient water (for instance, below generally accepted standards, or to water which is considered unsafe), water sources especially groundwater which once provided safe water becoming increasingly contaminated. This is particularly the case with increasing brackishness or salinity, largely as a result of increasing extraction. Water supplies which were being used sustainably are now deteriorating as extraction is exceeding replenishment by a wide margin. For instance, poor people are having to use up more time and energy or money to obtain water and could be using these resources more effectively to address their other poverty constraints. Additionally, poor people are having little capacity or power to improve their environment on a significant scale or, more specifically, to improve their access to water.

Could desalination significantly improve access for poor people living on marginalized land to improved water supplies? This possibility would need to be looked at with regard to the criteria outlined in the previous section. However, on three of the above aspects where there could be some impact would be on accessing cleaner and safer water, being able to continue to make use of deteriorating water resources or be able to make use of new resources such as seawater, and also having the possibility that the time and distance they take to obtain water being significantly reduced. Additionally by having a level of control over water provision, this could contribute towards empowerment and local people having greater decision making capacity over their own environment. Desalinated water can be used in different ways such as;

Drinking Water

The majority of output from desalination plants is used for drinking water. Some processes, such as distillation or reverse osmosis are used to produce almost pure distilled or de-ionized water. Others, such as electrodialysis can produce water which still contains some dissolved salts but which is, nevertheless, safe to drink. Most people do not like to drink distilled water, as it can reduce the taste of beverages made with it, and a small amount of some mineral salts dissolved in water is said to have beneficial health attributes. Hence it is usual to add a small amount of mineral salts to distilled water if it is to be used as drinking water. With some types of inland brackish or saline water, if this is otherwise clean, it might be possible to add a small amount of this to the distillate. With seawater and some water from inland sources chemical dosing might be required. Although animals would drink distilled water, it would also be better for them to drink water with a small dissolved content as well as this can be helpful to reduce illnesses.

Irrigation

Desalination is unlikely to be viable for irrigation purposes. This is because it is relatively expensive to treat water by desalination, most crops require large amounts of water for irrigation, and the price at which crops could be sold at would be unlikely to cover desalination costs. According to Commonwealth Science Council view points, it has been estimated, for example, that to irrigate an area of a quarter of a hectare typically twice a month would require the same amount of water as for supplying drinking water for 1000 people, 250 cattle and 500 sheep. However, irrigation from desalination might be viable if this is for high value crops and uses efficient, although expensive, techniques such as drip irrigation

Hydroponics

Hydroponics is the process of growing crops or plants without soil. Plants are grown in an inert mineral medium which is flushed periodically with water containing nutrients. Hydroponics is generally undertaken indoors and, sometimes, under artificial light. It is a suitable, although expensive growing technique in areas where water is scarce as it has been estimated that it requires about a tenth of the water for hydroponics compared with conventional farming of the same crop. Serious hydroponic farmers use distilled water as the base for making up the nutrient and mineral feed for plants, as then it is possible to determine the amount of minerals fed to the plants precisely. A number of proprietary small-scale desalination plants, generally based on reverse osmosis, have been developed specifically for the hydroponics market.

Vehicle Batteries

Distilled water used to be in high demand for topping up lead-acid batteries for vehicles. However, with the more widespread use of sealed for life batteries, it is not known if this demand is still as high. It used to be possible for a small entrepreneur with a solar still to make a reasonable income from making and selling distilled water for batteries. In Sudan in the 1980s, after researchers visited a pilot desalination project several years after it had been set up, they found that the project was no longer operating, but a garage owner had built and set up his own still for distilled water and apparently was barely able to keep up with demand for water for batteries.

Health Services

Distilled water is used by hospitals in a variety of applications, but often in blood transfusions and, laboratory testing of blood and cells, and for preparation of microscope specimens.

Laboratories

Distilled water has a variety of uses in laboratories, but especially in chemical analysis. Small-scale proprietary units for making distilled water for laboratories, generally based on reverse osmosis, have been developed.

Cosmetics

Distilled, or at least almost pure, water is specified in many cosmetic formulations to achieve products with the required characteristics.

Use of the Solid Residue

The solid product left behind after distillation processes might in some cases be too impure to be used for flavoring food for human consumption, or for industrial or chemical processes, though salt from seawater desalination would generally be acceptable for human consumption. The residue would be suitable in most cases for animals, for example as cattle lick.

POSSIBLE PROBLEMS WITH SEAWATER DESALINATION

There are several problems associated with desalination of seawater in different parts of the world. These include;

- Non-sustainable cost burden on users in maintaining the unit (whereby maintenance of the unit could be difficult to fit into the livelihood strategies of the community)
- Desalination could contribute to increase in the deterioration of a groundwater sources due to increasing extraction and increasing the salinity or brackishness of the groundwater. This would be doing little to enhance the long-term sustainability of maintaining a particular aquifer
- Desalination could also cause marginalization of some people
- Desalination can continue to maintain a short-term problem while obscuring a longer term solution. A particular scenario is that many of the poorest people in rural areas are dependent on shallow wells and many of these would be saline. The long-term solution could be a deeper borehole to tap into better reserves, improving rainwater catchment, or tapping into surface ponds, reservoirs or streams.
- Desalination might be continuing to make a deteriorating situation worse.

With regard to marginalized communities and environments, it could be most useful to consider an integrated or holistic viewpoint of the situation of poverty, marginalization, risk and environmental degradation of a particular environmentally related area. This would consider not just problems related to water, but also other issues such as farming, enterprises, health, other natural resources management, markets, community organization among others. By taking such a holistic view it would enable the underlying problems of the area to be better identified and solutions outlined. These might or might not bring out desalination as a possible course of action. On the other hand, if desalination is identified as priority then a stronger case would have been made for it than if the analysis had only been undertaken through considering only the water needs. The integrated approach is implicit in the Sustainable Livelihoods Framework, which could be particularly useful to applying in situations where water supply problems are significant.

ENVIRONMENTAL IMPACTS OF SEAWATER DESALINATION

Impacts on Coasts and the Mediterranean Sea

Desalination plants extract marine water and after a process of inverse osmosis, the desalinized water goes to the users, whereas the remainder (the brine) is unloaded in the sea. The brine spill can affect marine habitats and species, especially if the spill consists of very concentrated saline water that eventually rests on the seabed. Many aquatic species live in a narrow range of water salinity and changes in salinity (higher or lower) can severely affect plant and animal populations.

A clear example is the marine prairies of Posidonia whose growth is affected by salinity above 42 per thousand (grams per kilogram). All the communities associated with this area will be negatively affected through diminished oxygen generation, biomass and plant strength. Due to its weight, the brine can affect marine habitats depressions of the bottom of the sea, like a bucket, of sometimes 1 km2.

Energy Consumption

The energy consumption of the desalination plants is considered by different experts to be between 2.60 and 5 kwh/m^3. The estimations of concrete consumption vary widely according to the method of desalination used and the necessary pumping to elevate the water to the user level (assuming that most of the urban consumption is near the given coastline). Independent of the critical considerations of the need to desalinize an estimated 621 Hm^3 previously made.

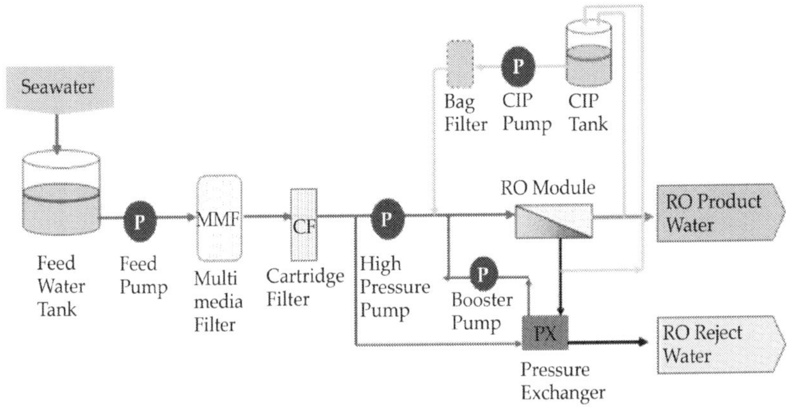

Figure 10.1: Power consumption through desalination process.

Source:*http://www.hitachi-aqt.com/products/standard_products/reverse_osmosis/sea_water/*

Research has established that the necessary energy for the desalination is very high (that is between 1,600 and 3,100 GWh/year) which means between 0.6% and 1.3% of the national electrical energy demand in 2003.

If this is balance energy demand with observing the Kyoto Protocol, and if it is considered that the desalination would take place with electricity coming from an actual mix (that is a composition of energy from hydropower, gas, or oil) and a resulting amount of pollution whose emissions are approximately 0.5 kg of CO_2 by kW hour, desalination emissions would mean between 0.8 and 1.5 million annual metric ton of CO_2 in the atmosphere additionally to that already emitted by the other industrial and automotive systems.

GLOBAL TRENDS ON SEAWATER DESALINATION

According to Commonwealth Science Council statistics it has been estimated that there are about 11,000 desalination plants in around 120 countries in the world. About 60% of these are in the Middle East. Approximate production is about 4 billion gallons (20 million cubic meters) daily. This would provide about 4% of the world's population with 75 liters per day, but taking all water consumption into account, including in agriculture and industry, this represents only 0.25% of the world's water needs. Most desalination plants process seawater rather than brackish water inland, though there are also a smaller number of inland plants, notably in India and the United States. Two thirds of the plants use vapors condensation technologies, while the remaining third are mostly based on reverse osmosis.

By far the largest producer of desalinated water is Saudi Arabia with about 5 million cubic meters per day. The majority of drinking water in Saudi Arabia is provided through desalination. The United States are also a significant producer, but of all the water consumed in the USA, that from desalination would be only a very small fraction. Although production is smaller in countries, mainly island states, such as Cyprus, Malta, Gibraltar, the Canary islands, Cape Verde Islands and the United Arab Emirates, it accounts for a significant proportion of the water used in those countries.

Small plants of less than 20 cubic meter capacity are likely to account for only a very small proportion of the world's desalination capacity. It is estimated that in the early 1990s there were just over 100 solar thermal plants at this level of operation, excluding an unknown number of very small units of several liter per day capacity, throughout the world. There would also be even fewer of this capacity operating on reverse osmosis or electrodialysis. Most of these are likely to be one-off operations, without significant institutional support or private sector capacity to supply plants of this scale in the commercial market.

The most significant obstacle to the development of small-scale plants is their cost relative to their level of output. They are also demanding on maintenance relative to their output. Small reverse osmosis units are used on boats, where the alternative of carrying large tanks of freshwater would also be costly, as well as inconvenient, but in terrestrial applications there are generally alternative sources of water supply in many areas.

RECOMMENDATIONS FOR DESALINATION PLANTS INSTALLATION

According to Commonwealth Science Council review points, there several recommendations that a government of specific countries can take into account when considering desalination plant installation. These recommendations include;

Estimation of Water Demands

Before beginning the construction of the new desalination plants project, a new analysis of the demands need to be made, considering the aspects mentioned previously: 'illegal consumption' and realistic estimates of consumption and recovery of costs through suitable tariffs. The measures of management of the demand included in the 'new culture of water' would have to be promoted among agricultural, urban and industrial users.

Environmental Evaluation and Assessment

A Strategic Environmental Assessment for the actions forecasted by specific government for the desalination plants should be made.

Temporary Planning

Necessary desalination plants, according to the estimates of demand and based on the Strategic Environmental Assessment, should be constructed gradually as much to adapt to the increases of demand as to take advantage of technological innovations.

Location of the Plant

Seawater desalination plants, like the facilities for energy generation, should be constructed in industrial areas of the coast, for example the harbor zones. Construction should not take place in natural areas, or in locations that can affect protected areas.

Power Supply

Desalination uses exclusively of new renewable power plants (mainly, thermoelectric and Aeolian).

Zero Spills

Respective governments should consider analysis options for other end-uses of brine instead of spilling them into the sea, thus obtaining desalination with zero spill requirements. The desalination process can generate other uses such as salt for aquariums or transfer to existing salt mines. In case spills cannot be avoided, it would be necessary to take a series of measures to diminish the effects of brine spills. For example, it would be important for the spill to dissolve as rapidly as possible. Total volume should be passed through diffusers in a fan, places at least one meter apart for proper diffusion and dilution of brine. It is important that the spill takes place in any case to level zero or seawater level, not the seabed, and never with emissions as some official documents suggest. Disposal should take place at points of the seabed with little slope to facilitate the dispersion of the brine and the dissolution of the salt. This is different to urban or industrial residual water spill where the blend of compounds would require the opposite physical conditions for disposal.

REFERENCES

1. American Water Works Association. (2011). Desalination of Seawater: M61. New York: American Water Works Association,.

2. Andrea Cipollina, Giorgio Micale, Lucio Rizzuti. (2009). Seawater Desalination: Conventional and Renewable Energy Processes. Berlin: Springer Science & Business Media.

3. Andrea Scozzari, Bouabid El Mansouri. (2011). Water Security in the Mediterranean Region: An International Evaluation of Management, Control, and Governance Approaches. London: Springer Science & Business Media.

4. Asano, T. (1998). Wastewater Reclamation and Reuse: Water Quality Management Library, Volume 10. Berlin: CRC Press.

5. AWWA Staff. (2011). Planning for the Distribution of Reclaimed Water, 3rd Ed. (M24). London: American Water Works Association.

6. Aysel T. Atimtay, S. K. (2011). Security of Industrial Water Sup-

ply and Management. Berlin: Springer Science & Business Media.

7. Brears, R. C. (2016). Urban Water Security. London: John Wiley & Sons.
8. Bruce Lankford, Karen Bakker, Mark Zeitoun, Declan Conway. (2013). Water Security: Principles, Perspectives and Practices. London: Routledge.
9. C. A. Brebbia, J. S. (2006). Water Pollution VIII: Modelling, Monitoring and Management. London: WIT Press.
10. Centre, W. A. (2006). Rainwater Harvesting Innovations in Response to Water Scarcity: The Lare Experience. Nairobi: World Agroforestry Centre.
11. Chris Binnie, Martin Kimber. (2008). Water Reuse. Chicago: IWA Publishing.
12. Claudia Pahl-Wostl, Anik Bhaduri, Joyeeta Gupta. (2016). Handbook on Water Security. New York: Edward Elgar Publishing.
13. Company, D. P. (1992). Guidelines for Water Reuse. Berlin: DIANE Publishing.
14. David Devlaeminck, Zafar Adeel, Robert Sandford. (2017). The Human Face of Water Security. London: Springer.
15. Diego J. Rodríguez; Matthijs Schuring; Nansia Constantinou; Amanda Goksu; Danielle A. García Ramírez; and Luisa Mimmi. (2014). Water Security for All: The Next Wave of Tools. London: World Bank Publications.
16. Doro-on, A. (2011). Risk Assessment for Water Infrastructure Safety and Security. London: CRC Press.
17. Gulbenkian Think Tank on Water and the Future of Humanity. (2013). Water and the Future of Humanity: Revisiting Water Security. Berlin: Springer Science & Business Media.
18. Headquarters, D. o. (1986). Water Desalination. London: Headquarters, Department of the Army.

19. Heather Kinkade-Levario. (2007). Design for Water: Rainwater Harvesting, Stormwater Catchment, and Alternate Water Reuse. London: New Society Publishers.

20. Initiative, T. W. (2012). Water Security: The Water-Food-Energy-Climate Nexus. New York: Island Press.

21. J. Anthony Jones, Trahel Vardanian, Christina Hakopian. (2009). Threats to Global Water Security. London: Springer Science & Business Media.

22. Jennie Barron, S. E. (2009). Rainwater Harvesting: A Lifeline for Human Well-being. London: UNEP/Earthprint.

23. K., A. S. (2005). Water Pollution. Chicago: APH Publishing.

24. Klinger, Jochen, Goldscheider, Nico, Hoetzl, Heinz. (2015). SMART - IWRM - Sustainable Management of Available Water Resources with Innovative Technologies - Integrated Water Resources Management in the Lower Jordan Rift Valley : Final Report Phase II. Berlin: KIT Scientific Publishing.

25. Kumar, A. (2004). Water Pollution. Delhi: APH Publishing.

26. Kurian, Mathew. (2014). Institutions for integrated water-resources management in river basins: An analytical framework. Chicago: IWMI.

27. Laosheng Wu, Weiping Chen, Christine French, Andrew C. Chang. (2009). Safe Application of Reclaimed Water Reuse in the Southwestern United States. New York: UCANR Publications.

28. National Research Council, D. o. (2012). Understanding Water Reuse: Potential for Expanding the Nation's Water Supply Through Reuse of Municipal Wastewater. London: National Academies Press.

29. National Research Council, Division on Earth and Life Studies, Commission on Geosciences, Environment and Resources, Committee on the Use of Treated Municipal Wastewater Effluents and Sludge in the Production of Crops for Human Consumption. (1996). Use of Reclaimed Water and Sludge in Food Crop Pro-

duction. New York: National Academies Press.

30. OECD. (2013). OECD Studies on Water Water Security for Better Lives. New York: OECD.
31. P. K. Goel. (2006). Water Pollution: Causes, Effects and Control. London: New Age International.
32. Qiang Zhu, John Gould, Yuanhong Li, Chengxiang Ma. (2005). Rainwater Harvesting for Agriculture and Water Supply. Tokyo: Springer.
33. Richard Helmer, Ivanildo Hespanhol. (2002). Water Pollution Control: A Guide to the Use of Water Quality Management Principles. Berlin: CRC Press.
34. Roberto Lenton, Mike Muller. (2012). Integrated Water Resources Management in Practice: Better Water Management for Development. Berlin: Routledge.
35. Schäfer, P. J. (2012). Human and Water Security in Israel and Jordan. London: Springer Science & Business Media.
36. Shimelis Gebriye Setegn, Maria Concepcion Donoso. (2015). Sustainability of Integrated Water Resources Management: Water Governance, Climate and Ecohydrology. New York: Springer.
37. Theib Y. Oweis, Dieter Prinz, Ahmed Y. Hachum. (2011). Rainwater Harvesting for Agriculture in the Dry Areas. Berlin: CRC Press.
38. U. Aswathanarayana. (2007). Food and Water Security. Berlin: CRC Press.
39. UNESCO. (2009). The United Nations World Water Development Report – N° 3 - 2009 – IWRM Implementation in Basins, Sub-basins and Aquifers (State of the Art Review). London: UNESCO.

INDEX

T

U

V

W